# OUR SENSES

# OUR SENSES

## AN IMMERSIVE EXPERIENCE

ROB DeSALLE

ILLUSTRATED BY
PATRICIA J. WYNNE

Yale UNIVERSITY PRESS

NEW HAVEN AND LONDON

Yale University Press books may be purchased in quantity for educational, business, or promotional use. For information, please e-mail sales.press@yale.edu (US office) or sales@yaleup.co.uk (UK office).

Designed by Nancy Ovedovitz. Set in type by Integrated Publishing Solutions, Grand Rapids, Michigan. Printed in the United States of America.

Library of Congress Control Number: 2017944414

ISBN 978-0-300-23019-2 (hardcover : alk. paper)

A catalogue record for this book is available from the British Library.

This paper meets the requirements of ANSI/NISO Z39.48-1992 (Permanence of Paper).

10 9 8 7 6 5 4 3 2 1

# CONTENTS

similar. Most organisms have a failsafe mechanism for too much heat, called heat-shock response. As part of this response, certain genes produce proteins that are activated when extremes of temperature are encountered. These proteins help the cells cope with the raised temperature. Other genes are present in organisms to help them cope with cold and are referred to as antifreeze proteins. These proteins also assist the cells in dealing with very cold temperatures. Yet, even though in these cases the molecules of cells "perceive" hot and cold, is this perception the same as takes place in a whole organism? Whether organisms perceive hot or cold depends on whether they have a brain and how that brain processes the information.

For example, certain genes in the fly genome can be mutated so that a fly will not perceive hot or cold. The experiment for detecting these mutants is brutal, in that flies are placed on hot plates and observed. A wild-type fly will skedaddle once the hot plate exceeds ten to twenty degrees above body temperature. Mutant flies, however, will sit until their legs begin to fry. Similar mutants for cold tolerance in flies have also been detected. We can infer that a fly perceives hot and cold, because a response is generated (the skedaddle). Bacteria more than likely do not perceive hot and cold, but their proteins do. So, there is an important line to draw when considering perception and simple physiological response.

Other senses not in the big five might include sensing time of day, magnetic and electric fields, changes in blood pressure, and hunger, among others. But we again need to determine whether we actually perceive these in a neural sense or whether our cells are simply responding physiologically to some external change. Consider blood pressure, for example. When blood pressure rises, we might or might not know it. An extreme rise in blood pressure will trigger certain responses in our bodies that our brains don't outwardly recognize. More than likely the response is physiological and preprogrammed and not perceived. In other words, your body recognizes or senses the change in blood pressure, but you do not always intellectualize it, nor do you often perceive it in the same way that you perceive light hitting the retina of your eyes. So, for the purposes of this book, I will consider these latter candidates as non-Aristotelian senses. I will examine some of them as part of evolutionary systems that arose as responses to environmental change, but when I discuss perception, I will stick mostly with the big five, along with balance, pain, and temperature.

Instead of taking the usual textbook approach of separating the five senses and explaining them in discrete chapters, I will use six important phenomena researchers have recognized that have a role in explaining the senses. First, our senses have arisen as a result of the evolutionary process, and we can learn a lot by placing our neural systems in this evolutionary context. Second, although we can perceive a pretty amazing broad swathe of our outer world quite well, human senses have limitations. But other organisms sometimes have super senses that are an excellent way to describe how they work and what the limits of the range of senses really are for humans. Third, within our species there is a great deal of variability—there are people with super senses and diminished senses. Supersensing and diminished-sensing humans not only provide a great way to describe the senses but also illustrate how we use the senses to interpret the outer world around us. Fourth, the senses of some humans have been altered because of trauma. Researchers have learned a lot about these injury-induced anomalies, especially how they relate to brain function. Our senses are also involved in the first steps in interpreting phenomena and situations we encounter. Fifth, our senses interact with one another to produce a coherent perception of the outer world. The extreme of this crossmodality is synesthesia, a situation in humans that mixes the senses so that synesthetes can taste colors and shapes, for instance. A discussion of crossmodality illustrates how our senses and perception work. How we perceive music, art, food, and other external stimuli will also help us understand our senses. This sixth approach to examining our senses illustrates how our senses communicate with one another and how our brains accomplish higher-order functions, such as reading or making music or producing art. In addition, how we place or map ourselves in the context of the outer world using these higher-order functions is an important step to understanding consciousness.

As an example, consider hallucinations. These are an anomalous aspect of consciousness, which are incredibly intense, altered, and sense-based experiences. These alterative perceptions of reality reveal new understanding of how our senses work. Auditory hallucinations are particularly fascinating and are often used in making a diagnosis of mental illness. But much can be learned from studying their origin and manifestation.

Sometimes the hallucinations can be problematic with respect to the adaptation of the individual to a normal life, but for many artists and musicians, auditory hallucinations have been a source of creativity. It is clear that structural and developmental aspects of the brain are involved in the phenomenon. In addition to hallucinations, culturally important endeavors such as music, art, and reading are a gateway to understanding consciousness. No two people will react to a famous work of art in the same way, the taste and texture of food, or the interpretation of music. The role of the senses in such variation is relevant. Furthermore, new research reveals how crossmodalities operate in enhancing musical ability and perhaps even appreciation of music or how vision, eye disease, and art interact to affect the creation of art.

But another reason exists for the difference in impact that a work of art or literature or a piece of music might have on a person, and that concerns the context within which the brain of the observer or hearer places the work of art. Each person has a set of memories, and the impulses derived from our experiences of and encounters with art, music, and literature are then placed into context by our brains. Teasing apart how our brains work in this dance of the senses with our experiences is an important part of understanding our humanness. For example, new research published in 2016 considers the important role emotion plays in how a jazz musician composes music.

Finally, what limits do the senses have? I conclude there are no limits to what we will be able to sense in the future, because humans are continually inventing methods for expanding the somewhat narrow ranges of our evolved senses of seeing, hearing, tasting, smelling, and touching. Focusing on the senses offers us the opportunity to delve into how we perceive ourselves as a species and how we perceive the outer natural world as conscious beings. Our senses lend to us a window into our very perception of reality and what consciousness means to us. All of our senses start with some outside influence—a flash of light, a bit of sound or a small molecule floating through the air, or a particle landing lightly or rushing onto the tongue. A signal produced by the external stimulus is transferred to the brain by nerve impulses, and in very specific regions in the brain, the stimulus is interpreted so that we perceive something from the outer world. Our memories then kick in, and we interpret the sensation in the context of our existence, of our past, and of our needs and wants.

I close with a discussion of the future of our senses and how consciousness might be altered. For instance, for the past two decades or so our visual, auditory, and tactile senses have been exposed to computer technology that has a huge impact on our day-to-day neural processes. Virtual reality only gets more and more real and more and more virtual. In addition, brain-computer interface developments have resulted in several technological advances for restoring hearing and sight to those individuals who have lost or never had those senses. *All* these modern advances have affected how our sense of reality and consciousness is formulated in our brains. I therefore hope you enjoy this neural and evolutionary romp through your senses!

# ACKNOWLEDGMENTS

I thank Jean Thomson Black for her unwavering editorial support and advice during the writing of this book and Michael Deneen for editorial assistance and organizational help. I thank Laura Jones Dooley for her meticulous copyediting and Margaret Otzel for her care and diligence during the production of the book. I also thank Leslie Nelson Bond for her careful reading and proofing of the proofs. Vivian Schwartz dutifully read and commented on earlier drafts, as she has done for several other projects with which I have been involved.

I am utterly indebted to the American Museum of Natural History's Exhibition Department for undertaking the development of the *Our Senses* exhibition that ran at the museum from November 2017 to June 2018. Specifically, I would like to thank Lauri Halderman, our Vice President of Exhibitions, whose amazing insight into subjects ranging from tornadoes to neurons, from black holes to wasp nests, and from microbes to ecosystems guided the development of the exhibition. I also thank Martin Schwabacher, Alexandra Nemecek, and Margaret Dornfeld, the writing team at the AMNH for the show, who so expertly translate the tough, sometimes esoteric science of the senses into something palatable for the public. Finally, I thank my friend and colleague Ian Tattersall for the many conversations (and beers) we have had over the past decade about science, evolution, and life in general.

# 1 THE BRAINLESS MAJORITY

*Sensing the Environment in Organisms without Brains*

*"Plants don't have a brain because they are not going anywhere."*
—Robert Sylwester, professor of education and philosopher

Our brain and our senses are the products of an experiment, billions of years old, that has occurred on our planet. Sorting out which events matter in that experiment for us to understand our unique capacities for perceiving the world around us requires an evolutionary approach. And that approach requires us to focus on a couple of important outcomes of evolution: biodiversity and exceptions. An amazing diversity of organisms have existed over the 3.5 billion years that organismal life has evolved on Earth. Without this diversity, we could not examine many nuances of our own sensory capacity. Our ability to hear sounds in a specific range, for example, is well described, but we would not know that our auditory range is biologically limited without our knowledge of how bats echolocate. So, an exploration of biodiversity puts our own biological characteristics involved in sensing into perspective. Exceptions in nature draw our attention to the nitty-gritty of how nature works and allow us to question why they occur. Examples of exceptional sensing come both from nature and from the record of human sensory limits. Many exceptions have evolved relative to our lineage. Some of these sensory exceptions help us understand how a particular sense works as well as how a sensory response might have evolved.

One example of this utility of sensory exceptions is how olfactory genes in animals are distributed. The number of functional olfactory genes found in vertebrates so far ranges from fewer than twenty in the green anole, a small lizard, to more than two thousand in the elephant. By comparison, humans have a respectable four hundred or so olfactory genes. If we couple these gene numbers with how different animals smell, we can learn a lot about our olfactory sense. The diversity of organisms on this planet reveals amazing natural experiments and offers great explanatory power with respect to understanding the senses.

All species are related through common ancestry, and this allows us to look at the steps that might be involved in the evolution of our unique sensory capacities. The tree of life is a superb way to demonstrate both the importance of biodiversity and the utility of common ancestry. For this reason, throughout this book I will use the tree of life as an organizing principle for our sensing organs and the organ that processes the sensory input (the brain).

The human brain, a most complex structure, is where our senses are processed and where perception exists. The brain evolved in higher animals to collect information from the outer world, to make sense of those data, and to promulgate survival. Most of the almost two million species that scientists have named and described to date have brains. (Scientists discriminate between the raw number of species that exist and the number of named species that are out there because they consider a species that has not been named or described as somewhat meaningless in an ordered world.) This number might lead one to think that we live in a very "brainy" world and, hence, one that is nicely tuned to the senses we're familiar with. But the grand majority of the life on this planet do not have brains, and so not having a brain has also been quite successful with respect to survival. Organisms without a brain can nonetheless sense and interpret the environment they live in quite well. It turns out that these organisms without brains are a neglected majority.

Most organisms are single celled and still unknown to science. Recent work on the human microbiome reveals that, on average, more than ten thousand kinds of bacteria live on and in our bodies, many undescribed and unnamed taxonomically. And that's just our bodies. When oceans and soil are examined, the number of bacterial species explodes. In the 1980s, the famous entomologist Terry Erwin suggested the stunning possibility that ten to a hundred times more species of organisms might live on Earth than the 1.5 million or so known at the time. Then scientists began to discover more and more novel species of bacteria. In 2009, microbiologist Rob Dunn theorized that there are at least one hundred million species of microbes (journalists called this Dunn's Provocation), which suggests that at least two hundred million species of organisms are living on Earth. Most of these species are microbial and thus lack brains. To add to this brainless majority, consider that 99.9 percent of all the organisms that have ever lived on the planet have gone extinct. Given that bacteria and single-celled organisms existed for probably two billion years before animals and plants emerged, this makes the estimated number of single-celled organisms even more stunning. Organisms with brains are and always have been an extreme minority, making Earth a pretty brainless planet.

So why the fuss about brains? A brain isn't required for perception. Galileo Galilei once wrote, "Before life came, especially higher forms of life, all was invisible and silent although the sun shone and the mountains toppled." Galileo's statement in retrospect means that before the bacterial mechanism for detecting light evolved, there was no perception of light as light and, hence, no light. The first organisms to evolve cellular mechanisms for detecting light metaphorically shouted, "Let there be light!" These first sensing organisms more than likely focused on one environmental stimulus, such as light, or on a specific kind of molecule floating around, or gravity, or magnetism.

Andriy Anishkin and colleagues theorize that the primordial sense was more than likely a response to mechanical stress on the lipid membrane surrounding a cell. In other words, any physical force that displaced the primordial membrane was the first external stimulus that cells learned to sense. Experiments reveal that force on the outer lipid membrane of a cell can result in conformational changes in the molecules that might be embedded in the membrane. Such changes in molecular conformation can act like switches in the embedded cells. If the molecules are squished or contorted, they will change shape, which could turn on or off other responses inside the cell. One common prevalent force that the outside environment enforces on a cell would be osmotic pressure caused by different salt concentrations inside and outside the cell.

Anishkin and his collaborators suggest that forces like osmotic pressure outside primordial cells might have been the first sensory experiences that enclosed cellular life experienced. Indeed, the phenomenon still exists in modern cells and points to an evolutionary frugality over the 3.5 billion years of life on Earth. When a structure or process in evolution is found to be adaptive, it lives on in its descendants as a result of natural selection. But another interesting possibility is that unrelated organisms rediscover the process or structure over and over again in evolutionary history. Examples of this latter kind of evolution, called analogy or convergence, abound. Wings are a good example of convergence, having arisen independently in birds, mammals, insects, and pterosaurs.

The answer to the question posed earlier, "Why brains?" then, is that primordial single-celled organisms had extremely limited capacities to sense more than single environmental inputs, which meant that these organisms had very limited perceptions of their environments. Brains evolved to allow for more precise integration of sensory input and for more exquisite perception of the information from the environment. Brains make our environment more understandable by detecting and processing a wider range of the outer world stimuli that are continuously bombarding us.

Enormous amounts of chaotic information stream, float, and dart around any organism's environment, confront its sensing organs, and have to be processed by a brain. One form of that chaos is best described as coming in waves. For the purpose of understanding how information from light enters our nervous system, we can say that electromagnetic radiation like light behaves both as a wave and as a particle. This means that light has qualities that waves and particles have. One characteristic of a wave is its length. Next time you are at the beach, watch the waves coming in. The wavelength is the distance from one wave's peak to the next one's peak. Electromagnetic radiation of different wavelengths (fig. 1.1) have different characteristics, and they can range from 0.00000000001 meters (gamma rays) to more than 10,000 meters (radio waves). Humans can detect light in only a very narrow range of this spectrum, from 400 to 700 nanometers, or 0.0000004 to 0.0000007 meters. The unseen (by human eyes) range of light outside the small end of the spectrum of wavelengths (400 nm) is what is called ultraviolet, or UV, light. Just outside the larger end of the spectrum

## BOX 1.1 | WHY AND HOW DO WE SEE COLORS?

When light hits an object, it slams into a large number of molecules that make up the object. Since light and electromagnetic radiation are also considered particulates, researchers have given the fundamental particle of electromagnetic radiation a name—the photon. When it runs into something, a photon has two options: it can be either absorbed or reflected. So, when light (which consists of photons of varying wavelengths) hits an object, millions and millions of interactions are taking place. Some molecules will reflect the photons, and others will absorb them. The photons that are reflected then hit our eyes, giving color to an object. For instance, plant tissues contain a molecule called chlorophyll. Because of its shape and size (it looks a little like Thor's hammer), this molecule absorbs light at 430 and 662 nanometers. These two wavelengths are where blue and red light, respectively, reside. So, chlorophyll does not absorb light between 430 and 662 nanometers, which is the wavelength for green light and a part of the color spectrum we see. The unabsorbed green light has only one place to go, and it is reflected off the plant. If a broad range of light hits an object that has ways of absorbing the different wavelengths, then the object absorbs all of those wavelengths. The object will have no photons bouncing off it in the visible range, and it will for all practical purposes have no color. The colors organisms detect, then, are simply the result of the reflection of light at different wavelengths to our eyes.

(700 nm) is infrared, or IR, light. In between are the colors we see from smaller wavelengths to larger—violet, blue, green, yellow, orange, and red. How and why our color perception got stuck in this narrow range of wavelengths is a story about evolution and adaptation. To understand this, we need to understand the physics of light and wavelengths.

We don't see the entire spectrum of light—say, into ultraviolet and infrared wavelength ranges and farther—because our eyes and the eyes of our ancestors evolved to detect only a narrow range of wavelengths. Although for most organisms the Sun is the major source of electromagnetic radiation, many other sources generate the photons that make up electromagnetic radiation. X-rays are an example of light created by the emission of electrons from atoms. Our eyes do not detect X-rays, but we have created a clever way of using photography to detect X-rays.

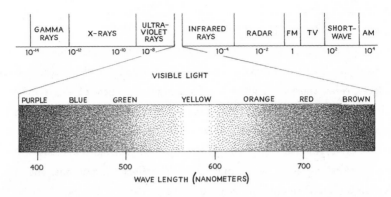

Figure 1.1. The range of photons in wavelength that we are exposed to in nature. Light waves range over eighteen or so orders of magnitude. The visible part of light is only a small sliver between 400 and 700 nanometers.

This theme of humans inventing clever ways of expanding the range beyond our natural limits, not only with seeing but with many of the other senses, is an important and ongoing concept. Other sources of wavelengths include bioluminescence. This form of light is emitted in our visible range and results in spectacular instances of living organisms producing and not reflecting light.

Another part of the chaos is molecules in the air as well as in the solids, gases, and liquids with which we come into contact. These molecules consist of atoms that form complexes in many ways, creating a plethora of incredibly small objects floating in the air or in what we ingest. Some of these molecules are very small, but all have distinctive shapes and sizes and can be detected as unique through a lock-and-key mechanism implemented by proteins embedded in the cell's membrane. Parts of these proteins flap continuously outside the cell and act as locks. When a small molecule comes along that fits the lock like a key, it forms a complex with the protein embedded in the membrane and changes the protein's shape. This change initiates a set of reactions inside the cell, causing a chain reaction that changes the state of the cell. What happens in the cell is called signal transduction, and this process is at the base of how our nervous system works as well as how single-celled organisms react to external stimulation. These small molecules floating around in our environment are the basis for how we and other organisms taste and smell.

Sometimes the displacement of the air (or water if we are swimming) around us causes sensation. Think of when we use a hand dryer in a

## BOX 1.2 | HOW DOES SOUND WORK?

Sounds are wavelength-based stimulations to our senses carried through water, air, gels, and other media as vibrations. Sound waves tend to displace air and other particles floating in the air. A range of sound exists because different sources can emit different wavelengths. As with light, organisms on this planet have evolved to detect sound waves in a narrow range relative to the overall range of sound. Sound waves travel in cycles that go from one wave peak to the next. The lower the number of cycles per unit of time, the lower the sound will be. The higher the number of cycles, the higher the sound will be. The unit for sound is called a hertz, and it measures the number of cycles per second of a sound wave. Humans can hear sound over a range of three magnitudes of hertz (from 20 Hz to 20,000 Hz), but other animals on the planet can hear sounds that are lower and higher.

public restroom: we can feel the air being displaced by the blower on our hands. We can also feel, as anyone can attest, when our head comes into contact with something solid, such as a low beam in the basement. So, when our skin comes into contact with a gaseous, liquid, or solid object, we experience a mechanical reaction. Organisms also need to know where they are in space, so many life-forms have evolved ways of keeping track of their position, and this leads to balance. The chaos of environmental stimulation that causes the need for balance comes from gravity and from the organism's movement. Other environmental variables include temperature, magnetic fields, and electrostatic fields.

Specialized cells in an organism detect sensory information from the environment, but how do they do this? The mechanism for single-celled organisms is very different from that of such multicellular organisms as plants and higher animals. In higher animals, brains process the information received from the sense organs.

Even those lineages of the single-celled organisms that we call bacteria and archaea can sense aspects of the world around them. This is because the environment comes into contact with things around these tiny organisms all the time. One need only view videos of predatory bacteria eating prey bacteria to realize that sensing is taking place. As the predators begin to decimate the prey, it's stunning how rapidly the prey bacteria selectively disappear. Here, a sensing of "You are my species, I'll leave you alone . . . and you aren't, so therefore you are good to eat," is being accomplished very efficiently. Even more impressive are videos of single-celled eukaryotes chasing and engulfing other single-celled organisms. But to me the most impressive video of bacteria sensing the external world is one showing "line-dancing" microbes that respond to magnetism as described below.

Some bacteria can count, and this capacity requires that the counting cell sense its surroundings. Quorum sensing is perhaps one of the most primitive ways cells sense and communicate with each other. But the basic theme of using molecules to communicate sense permeates all life on earth. Just as the quorum-sensing mechanism is based on molecular interactions, so are the senses of so-called more complex organisms. Single-celled organisms have a molecular system for detecting light, and some microbes (and indeed more complex animals) can sense magnetic fields. Magnetotactic bacteria orient themselves along the Earth's magnetic field because their cell membranes contain small particles of iron sulfide or magnetite (magnetosomes) encased in a membranelike organelle, and they line up in this encasement. Even though these tiny particles are lined up, more is needed for the line dancing to occur. The aligned magnetosomes within the bacteria are arranged in parallel, giving the bacteria a dipole characteristic and transforming them into tiny magnets with magnetic poles. Many species of bacteria are magnetotactic. The phenomenon apparently evolved just once in the evolutionary history of microbes, because most magnetotactic bacteria are in the phylum Proteobacteria and two other closely related phyla. In addition, the genes that modulate the construction of this organelle are found clustered in the genomes of magnetotactic bacteria, leading to two interesting aspects of the evolution of the phenomenon.

First, the tight clustering of the genes involved in making up the magnetosomes and the organelle that contains them indicates a single mechanism for the phenomenon (at least in the phyla where magnetotactic bacteria are found). Second, the genes are located on what is called a colonization island of DNA that can move around horizontally to other species. This horizontal transfer could implement and speed up the evolution of the magnetotactic trait in other bacteria.

In the absence of an interfering magnetic field, the bacteria align

## BOX 1.3 | QUORUM SENSING

Microbes often need to sense the density of their populations to respond to environmental challenges. The classic example of this kind of sensing, called quorum sensing, is found in *Aliivibrio fischeri*, a bioluminescent bacterium that resides in the light-producing organ of the Hawaiian bobtail squid (*Euprymna scolopes*). The squid and the bacteria have a mutualistic relationship whereby the squid cultures the *A. fischeri* while the bacteria light up the squid's light organ, which the squid uses to camouflage itself from predators. But the bacteria need to know when to light up, because it would be a terrible waste of energy to stay lit up all of the time. And so a mechanism for regulating the expression of bio-luminescent-producing proteins has evolved in the squid light organ that uses a clever capacity of the bacteria to sense their population size. The bacteria produce a protein called an inducer, which they recognize by another protein they produce called a receptor. When the inducer and the receptor bind to each other through (a lock-and-key mechanism), a cascade of genes in the bacterial genome is turned on, and a bioluminescent reaction occurs. When only a small number of *A. fischeri* are present in the light organ, the inducer is so dilute that it effectively does not bind the receptor, and no light is produced. This kind of sensing is entirely molecular.

with the Earth's magnetic field. Why would a microbe care about this? Because it craves nutrients for survival, and knowing the direction of this magnetic field helps it seek out more nutrient-rich environments suited to its biology. For instance, microbes in the phyla where magneto-tactic bacteria reside crave to be in a sweet spot in environments called the OAI, or the oxic (oxygen-rich)–anoxic (oxygen-absent) interface, as well as in anoxic areas very near to the OAI. They have evolved to prefer this environment, and because they have strong flagella that can propel them, they are on the lookout for such environments. It turns out that, because of the curvature of the Earth, not only does the Earth's magnetic field point north-south, but it also points at an angle to the surface. The angle to the Earth's surface allows orientation from surface to below surface. Being able to sense the Earth's magnetic field in the surface-to-below-surface direction allows magnetotactic bacteria to travel the most efficient path to the OAI, because the OAI is away from the surface.

What happens when the microbes are tricked into responding to magnetic fields other than the Earth's? Researchers at the Korean Institute of Science and Technology have built a tiny apparatus using a petri dish on a platform where a magnetic field is created below the dish. The magnetic device can be rotated with controls so that it overcomes the Earth's magnetic field and overtakes the behavior of the bacteria. Magnetotactic bacteria are placed into the petri dish, and the magnetic field is rotated with a dial by the "line-dance caller" in the lab—apparently to the tune of "Cotton-Eyed Joe." The magnetotactic bacteria respond with a decent version of line-dancing moves, by rotating in unison from left to right and forward to backward. The images of single-celled organisms dancing, and dancing well, are therefore quite humbling to a poor dancer like myself. Dancing is, of course, used as a metaphor in this instance, and it is important to recognize the metaphor for what it is.

Single cells need to know where they are in space and what they come into contact with and when. Since sunlight was a pervasive environmental factor billions of years ago, some bacteria also needed to know where light was and indeed used light as a means by which to live. So single cells developed fairly intricate and efficient ways of detecting external factors such as gravity, light, and environmental chemicals. Andriy Anishkin and his colleagues have proposed that tactile sensing in the original sense (as they call it) is a good argument for this being the first and perhaps most important sense a cell can have. But the order in which cells and organisms developed other senses would have to be speculation. We can, on the other hand, come up with pretty sound mechanisms as to why and how a cell might have developed a particular way of sensing.

Some bacteria use light as "food," just as plants do. One large group of bacteria that do this is the cyanobacteria. Molecules can interact with light by absorbing photons of light, and this is how these bacteria obtain light as a living. The mechanisms for using light as food in both plants and bacteria are practically the same. This observation at first glance is kind of wild. Bacteria and plants are not closely related, and there is no clear ancestor-descendant reason why plants should have a characteristic that bacteria have until one considers the origin of the chloroplast, which is the organelle in plants that converts light into nutrition for plant cells. The chloroplast in plants is actually the remnant of an ancient cyanobacterium that was engulfed by the ancestral plant cell. The symbiosis caused by the engulfed cyanobacteria in plant cells was such a lifestyle improvement for the ancestral plant cell that it stuck in an evolutionary context and is now a mainstay of plant life on the planet. The history of engulfment events by early eukaryotic cells of various kinds of bacteria is complex and sometimes convoluted. Some plant cells have engulfed other cells multiple times, and even multiply engulfed cells have been engulfed.

Another way that bacteria have exploited light is through altering the molecular properties of a class of molecules called opsins. These molecules are embedded in cell membranes where photons of light can hit them. Opsins have smaller molecules called chromophores latched to their inner structures. The chromophore clinging to the innards of the opsin forces the bigger molecule into a specific nonactive state while it lies embedded in the cell membrane. When light of a specific wavelength hits the cell, it also strikes the chromophore and causes it to be displaced, and the structure of the opsin itself changes, triggering other reactions in the cell.

In some single-celled bacteria, there is a molecule called rhodopsin embedded in the cell membrane that reacts with light. But unlike more complex organisms, the bacterial rhodopsin acts like a pump that brings high concentrations of chloride or moves protons into the cell that in turn changes the way the cell carries on its life. Single-celled eukaryotes also have rhodopsins that react when hit with light. The bacterial rhodopsin is pretty different from higher eukaryotic opsins, so whether or not the vertebrate opsins are derived from bacterial rhodopsin is not established. The point here is that the mechanisms for how opsins and rhodopsins detect light are similar and offer a preview to how higher animals sense light. Another point is that in single-celled organisms the mechanisms are carried out by proteins without a centralized need for processing the information in a brain. The "decisions" a single-celled organism makes as a result of environmental stimulus are rapid, chemical, and internal to the single cell. Higher organisms receive the environmental stimulation in ways very similar to single-celled organisms but do the processing of the subsequent information quite differently.

Multicellular life diverged from a single-celled ancestor about 1.5 billion years ago. A large number of single-celled eukaryotes exist today, and the patterns of their relatedness make it clear that there were many early events of divergence for single-celled eukaryotes into multicelled animals and plants. This observation holds because not all single-celled eukaryotes are each other's closest relatives and not all multicellular organisms come from the same common ancestor. Some single-celled eukaryotes, for example, are more closely related to plants than they are to other single-celled eukaryotes. For instance, the single-celled eukaryotes known as *Clamydomonas* (affectionately called Chlamy by people who study them) and algae are single-celled eukaryotes that are more closely related to plants than they are to other single-celled organisms like amoeba.

Plants can communicate the stimulus of the surrounding world to themselves quite well but have evolved very different mechanisms for doing this than animals. An excellent example is a sunflower: if you can, spend a few hours watching one respond to sunlight. Actually, the most interesting response the sunflower has to light occurs just before sunrise, when the plant's flower slowly turns toward where it anticipates the Sun will rise. The sunflower is pretty good at moving its floret and is very good at the timing. Another example is the mimosa, a plant that responds very rapidly to touch, and anyone who has seen *Little Shop of Horrors* should be immediately reminded of a Venus flytrap, which responds rapidly and voraciously to prey items that unwittingly wander across its trapping apparatus (fig. 1.2). Plants, however, do not have nerve cells and hence do not have a brain or a nervous system like animals. (I make these statements about plants and nervous systems even though a journal called *Plant Neurobiology* exists and even though several institutes are dedicated to the study of the neurobiology of plants. The focus of plant neurobiology is not the same as animal neurobiology.)

Metaphor has become important in the comparison of how organisms respond to the environment. An organism with a "metaphorical brain" like a plant does not process information from the external world the same way vertebrates do, but this is not surprising. By metaphorical brain, I mean a system that is analogous to a vertebrate brain but not at all neural. This capacity to respond to the outside world is what prompted some researchers to initiate the plant neurobiology approach. But it is very difficult to deny that plants don't have brains and they don't have nervous systems. I prefer to acknowledge that plants are pretty good at

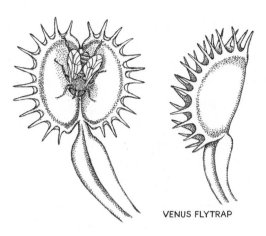

VENUS FLYTRAP

Figure 1.2. An example of plant neurobiology or plant intelligent behavior? The Venus flytrap (*Dionaea muscipula*).

sensing the outer world and have some way of centralizing their sensing of the outer world, but in a functional structural context, they do not have brains. In evolutionary biology, we might say that the plant version of a nervous system has converged on the insect or vertebrate brain. The plant's central sensing system is a metaphor for the invertebrate or vertebrate nervous system. It is intellectually much more pleasing to me to realize that plants have figured out a novel way to perceive the outer world that has nothing to do with a nervous system. And indeed, when we start to examine the ways that animals with nervous systems have evolved structures and mechanisms for the traditional senses, this theme is repeatedly borne out. So as Michael Pollan, the outspoken defender of plant life on this planet, suggests, perhaps we should call it, not "plant neurobiology," but rather "plant intelligent behavior." And in this context, plants have evolved intelligent behavior without any reference or evolutionary similarity to the animal way of getting at intelligent behavior, other than using some of the very basic molecular tools in the evolutionary toolbox that most multicellular eukaryotes have. This intelligent behavior allows the plant to sense stimuli from the environment such as light or chemical concentrations and to interpret them in an "intelligent" manner. The neural basis of plant intelligent behavior is simply another solution to cell-to-cell communication that life on Earth has discovered and evolved as a response to the need for sensory connection to the outer world.

It is no surprise that organisms without eyes, ears, noses, skin, and mouths can't see, hear, smell, touch, and taste. These so-called traditional or Aristotelian senses are the bailiwick of advanced animals. And so organisms without these attributes have focused their sensing on other environmental stimuli, like electrical and magnetic fields and chemical signals that don't behave like taste and smell. Organisms that can see, hear, smell, touch, and taste have evolved an amazing array of anatomical and physiological traits that enhance those senses. The breadth of mechanisms that life has evolved to sense the outer world of environmental stimuli is stunning.

## 2  BRAINS AND PRAINS

*Brains (or Not) in Animals from Sponges to Us*

*"A sponge sees everything? A sponge sees nothing."*
—Lawrence Tierney, actor

Specific sensory milestones arose as the tree of life's branches and twigs spread out. The very earliest milestones include the evolution of cell-to-cell communication through molecular processes such as quorum sensing. These milestones also included the evolution of sets of genes that perform specific functions in cells called molecular toolboxes for the transfer of information through single cells. Multicellular organisms that evolved later communicated between cells using processes like signal transduction. Once multicellularity arose, more options for how cells communicate opened up, and this development heavily influenced the biology of our direct eukaryotic ancestors. These milestones include action potential, synapses, differentiated nerve cells, neural webs, neural nuclei, and specialized nervous systems. And although these milestones, all reached in the evolution of our lineage, might seem like a streak toward perfection, they are mere stopping points. Other lineages were evolving, too, and our neural milestones more than likely mean nothing to these other lineages.

Charles Darwin likened organismal relations to what he called "the great tree of life" to represent the interconnectedness of all organisms on the planet through common ancestry and divergence. The great tree of life has become a common tool for biologists to trace the evolution of traits. By following the divergence events of organisms and of the traits that might be considered neural, we can get a pretty precise view of how brains in general, and our brain specifically, evolved.

The very early branches in the animal part of the tree of life include sponges and a small, pancake-shaped animal called placozoa. Some researchers think that all sponges are related to one another through a common ancestor to the exclusion of all other animals. Others suggest that there are two great lineages of sponges. What is certain is that sponges have only eight or so cell types, and none of them are neural, so they do not have brains. Placozoa, for their part, have four cell types, and none of them are neural, so they likewise cannot have brains. But what is interesting about these two early branching animals is that they have many of the genes necessary for making a neural cell but simply don't use them for neural cells. Apparently, they have found other uses for the genes that the rest of animals use to make nerve cells. Both sponges and placozoa can sense environmental changes and respond to them, making these organisms sensing animals without brains or a nervous system. Sponges can "sneeze" by sensing particles that they come into contact with. Their sneezes are spectacular events, in which the sneeze builds up over a period of an hour or so and needs to be viewed in time lapse. Placozoa can forage for nourishment and are very efficient at detecting food, and this without a brain.

Comb jellies, or ctenophores, are incredibly interesting animals. They look darn cool, and they have been proposed as the first animal group to diverge from the single-celled eukaryotic ancestor of all animals. They have several cell types, one of them neural, as well as a nervous system, but no centralized set of cells that forms a brain. If ctenophores are the first animal group to branch off the animal tree of life, then this opens some very interesting possibilities. One is that nervous systems are convergent in comb jellies and other animals (except for sponges and placozoa). The alternative is that nervous systems can be lost relatively easily, such as in the lineage that gave rise to sponges and placozoa. This example shows how using the concept of a common ancestor and a tree of life allows us to dissect the potential major changes

## BOX 2.1 | ACTION POTENTIAL

The physiological response of cells where the electrical charge rises and falls in a cell is called an action potential. More complex animals will maintain an electrical difference between the environments that are external and internal to their cells. A cell membrane regulates the difference. The membrane keeps the voltage inside the cell at minus 70 millivolts relative to the outside, which is called the resting state. When the cell is stimulated in a specific way (see text), the voltage difference outside and inside the cell rises by about 110 millivolts, so there is a new difference of plus 40 millivolts. The charge inside the cell rises as a result. The voltage then falls rapidly to well below its resting state before settling back to its resting state of 70 millivolts. The rise and fall of an action potential occurs in a very consistent manner and is how nerve impulses or electrical signals get transmitted from cell to cell throughout the nervous system. Action potentials are also important in communication of the nervous system with the sensory and motor systems.

of the nervous system that occurred in the divergence of animal nervous systems.

But the way neural cell communication works is common to these lower "squishy" animals and our lineage. The transfer of an electrical charge through action potential seems to be an ancient animal invention.

After ctenophores, sponges, and placozoa branched off, the common ancestor of cnidarians and the group we belong to, called bilateria, diverged. Any organism that has symmetry across a midline drawn down the center of the head to the end of the body is called a bilaterian. The group most closely related to bilateria are the cnidarians, which includes jellyfish, hydroids, corals, and an odd group of organisms called cubozoans or box-jellies because of their boxlike shape. Cnidarians have cells that can be called neural cells and a nervous system best described as a net, but no centralized set of nerve cells that could be called a brain.

Bilateria are divided into two big groups: the protostomes (for example, insects and mollusks) and the deuterostomes (which includes humans). Brains are found in organisms in both major groups of bilateria, so scientists generally consider the brain a bilaterian invention. Most biologists would describe the protostome and deuterostome brains as the same by using the term "homologous." Darwin first made this meaningful by pointing out that homologous traits are those that occur as a result of direct common ancestry. So bird wings and bat wings are not homologous, because there are many mammals without wings that disrupt the common ancestry of the two kinds of vertebrates. Any traits present in organisms that are not connected by direct common ancestry are considered "analogous" and have arisen as a result of convergence.

But is the brain really a bilaterian invention? The overall structure of protostome brains can be argued to be structurally different from advanced deuterostome brains. The common ancestor of bilateria had a neural network that reached most parts of its body, and in the anterior region, a patch of neural cells that neurobiologists would call a brain existed. Some neuroanatomists suggest that the common ancestor of bilateria even had a complex structure with three parts to it.

It is possible, though, that the patch of neural cells at the anterior end of the neural tract that traversed the body in the ancestral bilaterian took two different pathways when deuterostomes and protostomes diverged. I like to call the protostome brain a "prain" to differentiate it from what looks to me like a differently evolved and structured deuterostome brain. Although we can use insects as model systems to decipher how sensory information from the outer world is collected and turned into nerve impulses, once the information hits the brain in the case of deuterostomes and the prain in the case of protostomes, the similarities between these two types of neural centers cease. The only satisfying comparisons more than likely stop with the collection of the information by the sense organs and its transfer to the brain or prain, where the information is processed. More than likely, many of the tools in the sensory toolbox existed in the bilaterian ancestor. Since the bilateria diverged from each other these tools were being used diversely as speciation event after speciation event occurred. To be fair, some researchers argue for homology on the basis of another important piece of information. Protostomes like the fruit fly (*Drosophila melanogaster*) use many of the same genes for neural processing that deuterostomes do, and they also have very similar (but not identical) structures where nerve cells come into contact with each other (called synapses—see box 2.2). But it is an interesting idea to consider that complex brains have evolved

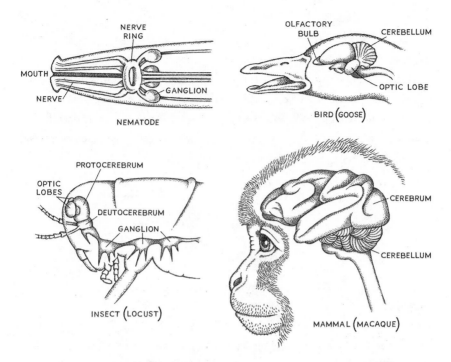

Figure 2.1. Comparison of vertebrate brains (bird and primate) with nematode and insect brains.

at least twice on this planet—once in the protostome lineage and once in the deuterostome lineage (fig. 2.1).

If we avoid calling the protostome brain the same kind of brain as ours, or at least not the same triune brain, then we can use the tripartite brain as a heuristic device to make some sense of how the vertebrate brain evolved. The early branching deuterostomes such as the starfish and sea urchins have neural cells, but they are distributed throughout the body with no central cluster of cells to call a brain. Some scientists split hairs and suggest that these organisms, also known as echinoderms, have what is called a distributed brain.

The next group of animals to consider as part of the context for hanging the human brain on the tree of life is a sister group to the echinoderms, the chordates, or phylum Chordata. All organisms in this group have nerve chords. The very earliest branching members are the urochordates and cephalochordates. Urochordates are strange animals that are best represented by the sea squirt, which is placed in a group called tunicates (because they form a tunic or coat of material similar

## BOX 2.2 | SYNAPSES

Synapses are contact points that permit one neuron to communicate with another by passing along an electrical signal. The mechanism by which the electric signal occurs is complex (see box 2.1). The cell delivering the electrical charge is called the presynaptic cell, and the one receiving is the postsynaptic cell. Researchers also include the communication from other cell types (such as muscle) to a neuron as being accomplished through synapses.

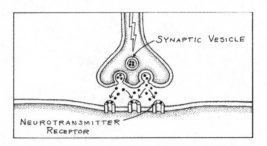

Figure 2.2. The intracellular region of the synapse. The presynaptic cell contains the synaptic vesicle that releases the neurotransmitters, and the postsynaptic cell has the neuroreceptor proteins embedded in the membrane.

to cellulose to protect themselves as adults). The sea squirt has a larval stage just like some amphibians. In its larval stage, it has what is called a cerebral vesicle, which is a technical way of saying that it has a place for a brain but doesn't fill it with a brain. Oddly enough, as it develops into an adult, the sea squirt larva feasts on its brain. As a larva it has no external mouth to push food through, and it has to get its nutrition from somewhere. Because as an adult it doesn't use the tiny clump of nerve cells in the cerebral vesicle, it reabsorbs the tissue so that when it transforms into an adult, it has no brain. It does have what is called a notochord (but no vertebral column—that comes in later in the tree of life). Cephalochordates also have a nerve chord, but again no backbone at all. One of the best representatives of cephalochordates is an organism called the lancelet. It is difficult to make a definitive statement about the existence of a brain in this organism, and in fact anatomists call what could be considered its brain a blister. The blister is literally that, a section of the notochord in the head region that has puffed out a little.

Vertebrates have a backbone (or spine) and fully developed nerve chord. Neurobiologists have had a hard time accepting that the vertebrate brain evolved linearly with simple additions of more complex parts. The limbic system and the cortex are two examples of more complex neural real estate that gradually occurred in common ancestors as animals diverged. One would think that more complex structures or processes should evolve from less complex ones. But it is not necessarily true that descendants are always more complex than their ancestors, and it is not true at all that the result of evolution is more complexity. Yet when the starting point is a primitive structure, the only way for it to evolve is to add on. And it seems that the vertebrate brain has evolved by adding layers. Uncovering how these layers were added and when is the problem. Schemes of brain evolution almost always involve accommodating three major brain parts, leading to the image, or heuristic, of a tripartite brain. There are, under some criteria, three major parts of the brain: the stem and cerebellum (also known as the R-complex—"R" for "reptile"), the limbic system, and the cortex. But these divisions are somewhat arbitrary, and this is more than likely why some neuroscientists reject the tripartite sectioning of the brain. However, as a heuristic, using a tripartite model of the brain for vertebrates works fairly well. Even with this disclaimer, there is still controversy as to how the tripartite brain evolved or even whether it is tripartite.

The ancestor of all vertebrates had a very primitive-looking brain with no or very little cortex and a very rudimentary cerebellum. Most of this ancestor's control of its behavior resided in the stem of the brain and the cerebellum, where very basic bodily functions are controlled, such as heartbeat and breathing. Some basic capacity to respond to the environment also existed in this ancestor's brain, such as sensing smells, sights, and sounds. This ancestor had what is referred to in the classical tripartite scheme as a reptilian brain. In the classical tripartite brain scheme, the final two layers added are the paleomammalian brain and the neomammalian brain. Most of us who work in evolutionary biology would infer from these names that the next two most important common ancestors, then, are long-extinct mammals (the paleomammal and the neomammal). The problem with this scheme is that it omits birds, which have pretty sophisticated neural ways of dealing with the world and hence brains that are hard to place in the continuum. There is therefore a need to either revise the names of these ancestral brains or assume as we have before that the bird brain that lies beyond the lizard part of the brain is not the same as the mammalian brain. We could rename the reptile brain our inner fish brain and try to move outward on the tree of life. We could also suggest that something like a reptile/bird brain came next, and finally the proto-neo-mammalian brain. The problem with this way of thinking is that fish are not a real group of organisms to some evolutionary biologists. Why? Although the name "fish" is descriptive, it does not define a group of organisms to the exclusion of all others, which is the way taxonomists usually name. The solution to this problem is either to name the major lineages of fish something else or to name all descendants of the common ancestor of what we call fish, fish. And this would include humans as fish. At some point, we need to know when to cut bait and when to fish. So let's cut bait on this scheme and look at the triune brain from a different perspective.

If one were to review all the major names that neurobiologists have given these three parts of the brain, one would encounter both a certain arbitrariness and a high degree of subjectivity. All are dependent on which species have which structures. What is clear from this exercise is that many scientists really like the idea of there being three major parts of the brain (fig. 2.3).

Consider the divergence of vertebrates and how this might add some objectivity to the discussion. After the major anatomical structures at the base of the brain evolved in the common ancestor of fish, amphibians, reptiles, birds, and mammals (the inner fish brain), the next major common ancestor that evolved was one giving rise to all higher vertebrates (reptiles, birds, and mammals), and this ancestor added a layer of the brain to accommodate more complex ways of processing the signals from the senses gleaned from the environment (our reptile brain). This led to the development of an inner region of the brain that includes what is commonly called the limbic system. This way of thinking is a reasonable way to keep a triune brain scheme alive, since all of the abovementioned vertebrates have a limbic system. The limbic system then started to undertake some very interesting tasks with respect how this ancestor reacted to the environment. The final addition was the cortex, which expanded in various ways in the descendants of this common ancestor. It is significant that the neocortex changes in specific ways in the descendants

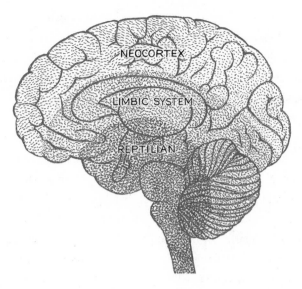

Figure 2.3. The triune brain.

of this common ancestor, as evidenced by the expansion of the neocortex in birds and mammals and the lack of expansion in reptiles. In short, the neocortex of birds is a different kind of cortex expansion than the neocortex expansion in mammals.

What this all means is that a complex limbic system probably arose in the common ancestor of birds, reptiles, and mammals. But so did a tiny primordial cortex. It is difficult to tease apart the possibility of two ancestors (an ancestor with a tiny primordial cortex and no limbic system or an ancestor without a tiny cortex and a limbic system), and so even though the function and anatomy of the brain appear tripartite, the additions cannot be interpreted as sequential. Part of the problem resides in how birds are related to other higher vertebrates. And again, birds are probably why neurobiologists who disdain the layering tripartite brain scheme do so. But two other kinds of animals also disturb this scheme: turtles and amphibians. Turtles have a thalamic reticular nucleus, a clump of nerve cells around a part of the limbic system called the thalamus, as do other organisms with a limbic system, but it is not as well defined, nor does it do the same things as the bird or mammalian thalamic reticular nuclei. The other critical class, amphibians, has some structures reminiscent of a limbic system, but they are not as well partitioned as even the most primitive reptile limbic systems. To summarize, the heuristic value of a triune brain is important. So, do we stop thinking in threes, or do we use the tripartite brain as a nice device to make sense of a very complex process? I think as long as we realize the limitations of the tripartite brain and don't try to force analogous structures as being homologous, then the pathway of using the tripartite brain as a heuristic is a good path to go down.

How did a complex brain that integrates sensory information from the outer world evolve? Modern biologists are careful not to attribute too much of what we see in nature as having evolved through natural selection. Richard Lewontin and Stephen Jay Gould in the late 1970s pointed out that much of evolutionary biology, notably human evolutionary biology, consisted of "just-so stories." They named this the Panglossian Paradigm, after Dr. Pangloss in Voltaire's novel *Candide*. Dr. Pangloss saw purpose in everything ("All is for the best"), and so the adaptationist program that Lewontin and Gould saw was so prevalent in biology at the time was aptly named.

I want to simplify the potential of natural selection in the evolution of the integration of senses with perception. But keep in mind that not all traits we see in nature are the result of natural selection. In addition, evolution does not strive toward perfection, as the ranges of our senses suggest. Quite to the contrary, the evolutionary process simply finds the best solution it can to an environmental challenge, and hence many of the characteristics we see in nature are not perfect solutions but rather stopgaps that rapidly and efficiently solve a challenge from the environment.

Let's think in threes one more time and oversimplify the environmental encounters organisms have into three basic categories that all organisms deal with. The idea comes from the fertile mind of a paleontologist friend of mine. He was a young graduate student taking a course from me while I was at Yale University as an assistant professor. In his simplistic way of thinking he came up with a probably more than oversimplistic scheme for how animals survive the challenges of the natural world. His scheme started with the animal coming into contact with another organism. This prompts the observer animal to place the intruder organism into one of three bins that will enhance or increase the fitness of the species. The first is easy. If the intruder is a fellow species member of the opposite sex, then it is placed into the "mate bin" (as in, "I mate with that"). If it is something that the observer judges as dangerous, then the intruder usually goes into the "run-away bin" (as in, "I run away from that"). The last bin is the "eat bin" (as in "I eat that").

Now, of course, an organism will refine its interpretation of what goes into these bins as it gains experience. For instance, our *Homo sapiens* ancestors altered their criteria for the mate bin so that it didn't matter if the intruder was the same species. If the intruder had two legs and stood upright, then it went into the mate bin, as indicated by genome sequencing evidence indicating that our species possibly interbred with archaic humans like Neanderthals. The run-away bin is probably the trickiest, because intruders larger than the observer often aren't dangerous and may even be beneficial, and those smaller than the observer often can be as dangerous, or more so, than large, vicious intruders. In addition, even if the interloper is of the same species, the observer might still want to run away from it. The last bin, the eat bin, is also a difficult one, because actually smelling or tasting the interloper, or at the very least getting a good look at it, is often required to make a decision. The faster and more efficiently organisms place other living creatures in their immediate environment into these three bins, the more likely they are to survive and to pass their genes to the next generation. Populations evolve, but individuals do not (individuals live and die, but they do not evolve). The fact that populations evolve means that for the three categories of binning experiences, there is variation in a population in how the members of the population categorize. Variation is the stuff that natural selection works on. The variation simply comes from the genetic makeup of the population where mutations can arise every so often that produce the variation. The variation would appear in the precision with which organisms encountered are binned and the amount of time it takes to bin and react to a challenge. If an individual in a population is imprecise or slow in binning something, it loses! It is gone, and so are the genes it could have passed to the next generation's population.

Although the preceding is overly simplistic, the three bins are probably real in some basic way. Of course, there are other bins, and there is leakage from one bin into another. But the point is to make clear that our brains are continually making interpretations such as the purported binning of other organisms I describe (including encounters with plants and microbes), and they have evolved to do it very rapidly. In fact, slight changes in the reaction time to some of the binning decisions organisms make can be the difference between surviving or not. This overly simple example also might make it look like natural selection is the only thing at work in how nature works. As discussed earlier, it is important to avoid the Panglossian view of nature; chance plays a huge role in how evolution works and indeed how these three bins might evolve in different species.

Charles Darwin focused most of his intellectual energy on natural selection, and thankfully he did, because he was the first (along with Alfred Russel Wallace) to come up with a mechanism by which evolution could work. He even called his 1859 book *On the Origin of Species,* one long argument for the existence of evolution by natural selection. This view of how life on the planet evolved was a pervasive part of the evolutionary paradigm until the 1960s and 1970s, when Motoo Kimura and others began to infuse into evolutionary biology that random factors are also involved in driving organismal change on the planet. This force of nature is called genetic drift.

The idea behind genetic drift comes from probability theory and suggests that evolutionary processes are like sampling problems. In any sampling process, there is a finite probability that a biased sample will result. In most cases, the finite probability is tiny compared to the force of natural selection, but in some cases sampling drift, as it is called, can occur with high probability. These cases occur when populations are very small. Think of it this way. If I were to bet you that I could flip one hundred heads in row with a fair coin, it would be a terrible bet for me almost every time I would make it. The probability of my flipping one hundred heads in row, while finite, is very small. But if I bet you I could flip a coin and get two heads in a row, that is a much, much better bet for me. Researchers easily started to see genetic or sampling drift in numerous natural cases, and its impact on how organisms evolve became integrated with ease. It is now thought that sampling drift and natural selection work in concert to mold genetic and phenotypic variation in nature.

The senses and brains of organisms have not escaped the forces of selection and drift. Moreover, they have been molded by these forces— so much so, in fact, that to leave both drivers of evolution out of the picture would be to miss most of the story.

## 3  THE MONKEY'S UNCULUS

*Tactile and Balance Sensory Capacity in Animals*

*"Nothing we use or hear or touch can be expressed in words that equal what we are given by the senses."*
—Hannah Arendt, philosopher

As the fetal brain develops in mammals, an incredible number of brain cells are produced every minute. In humans, that number is 250,000 cells, such that by the time a baby is born, there are more than 100 billion cells in its brain. How these brain cells are arranged and how they enervate the rest of our bodies is an essential part of the story of the senses. It is significant that most of the human brain is dedicated to bringing in and processing sensory information. How these nerve cells of the brain are wired and what they do depends on how the brain develops. Where certain sensory functions are arranged in the brain is a matter of mapping them. And how they are mapped is a fascinating story.

Brain researchers scrape for every research trick they can get. Part of the reason for this scrappiness is that observations of the brain's structure are best facilitated when the brain is outside the skull, and this requirement all but eliminated studying an active brain in this way in the early days of brain research. One researcher who overcame this limitation took an ingenious approach to understanding the localized function of the brain in patients undergoing brain surgery. Two small regions in the brain, called the sensory and motor strips, were the focus of Dr. Wilder Penfield. Penfield died long before the movie *Hannibal* was released, so he never had the chance to see it, but he did know, as Dr. Lecter says in the penultimate scene of the film, "You see, the *brain* itself feels no pain, if that concerns you, Clarice." And Lecter then demonstrates this cool brain fact by feeding the not-so-lovable Agent Krendler parts of his own brain. Penfield used this knowledge of the lack of sense of pain of the brain to accomplish a tour-de-force study in brain science.

Penfield was a brain surgeon, and before he would perform the actual surgery, when a patient's skull was opened and the sensory and motor strips were visible and accessible to touch, he would mechanically massage the strips in a sequential manner. Because the patient was every bit as awake as Krendler was during that scene in *Hannibal,* Penfield was able to ask the patient what happened after the stimulation. So, for instance, if he stimulated a certain part of the sensory strip, the patient might say, "I feel a tingling in my fingers." Penfield kept painstaking notes on more than 120 patients that he examined in this way (over his career he examined more than 1,000 patients using this approach), and in the end he was able to map both of these neural strips quite exquisitely and accurately. "Map" is a perfectly apt name for what he did, but the way he visualized the products of his brain tickling is somewhat nightmarish. By estimating the amount of neural tissue dedicated to sensing in one case and to controlling motor activity in another, he was able to commission an artist, Mrs. H. P. Cantlie, who in 1937 started to draw very strange-looking beings called homunculi (the plural of the word "homunculus").

A homunculus for the sensory cortex would look a bit different from a homunculus for the motor strip, because our motor and sensory abilities do not directly mirror each other. We use different amounts of the cortex to control how our fingers move compared with what we use to sense the outer world with them. It turns out, though, that Cantlie's drawings were often inaccurate. Psychologist Richard Griggs points out that Cantlie's drawings sometimes indicate that they are based on data for the left hemisphere, but the drawings of the homunculus for these data are for the left side of the homunculus, which is a mistake because functionally the left side of the brain controls the right side of the body. Griggs also notes that even though Cantlie never drew a female homunculus, in the late 1980s a female left breast started to appear in the classic drawings. The appearance of this left breast only partially corrects the bias of Penfield's homunculus toward male subjects. In fact, only about one-tenth of the patients he examined were specifically noted to be female. In 2012, four female neuroscientists discussed this sex bias and the lack of a *her*munculus in the literature. Their point was that males and females have some extreme differences in their sensory makeup that should be appreciated. Even with this sex bias, we can say that humans have a bizarre-looking sensory homunculus. Because the data are derived mainly from males, I will use the male pronoun forms. His hands look huge, his lips and tongue are supersized, and his penis is, well, really gigantic. He has extremely large feet and a torso that looks emaciated. This strange-looking being indicates that a large proportion of the sensory strip in the human brain is dedicated to our digits, lips, and genitalia (a *her*munculus's genitalia would be about as big as her real foot).

Since the late 1940s so many mammal homunculi have been described that they could fill a small petting zoo—platypus, rabbit, shrew, bats, mice, cat, dog, and monkey, for example (fig. 3.1). Ratunculus, platypunculus, and simunculus (rat, platypus, and monkey, respectively) are just a few of the strange names given to the even stranger-looking drawings. Perhaps the most revolting of all of the supersensed-unculi are those of mammals that burrow in the ground or that use the sensory apparatus of their snouts to interpret the world. The star-nosed mole is a bizarre-looking mammal in person, let alone as a molunculus. This burrowing mammal's snout has twenty-two projections (eleven on each side) that radiate outward and are numbered 1 through 11, with the eleventh being a small ray at the bottom of the apparatus. As this mole moves along in a burrow, the rays are continually reaching out and touching objects it encounters. It is estimated that the rays can make more than a dozen "touches" in a second. Two unique structural aspects of the arrays ensure that this mole is deeply in touch with its surroundings. First, each ray has hundreds of smaller sensory structures called Eimer's organs. These structures are distributed all around the ray and are enervated through a structure called tactile fovea. (In the case of sight, the fovea of

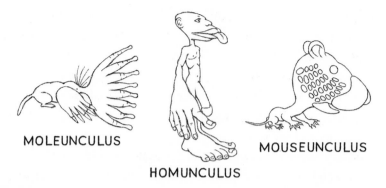

MOLEUNCULUS

HOMUNCULUS

MOUSEUNCULUS

Figure 3.1. Moleunculus (star-nosed mole), homunculus, and mouseunculus. The patches of cells in the mouseunculus head region are called barrels.

the eye act to sharpen focus in the center of our visual field—see Chapter 8.) Ray number 11 is actually part of the tactile fovea. Whenever rays 1 through 10 encounter something really interesting to the mole, ray 11 kicks into action and explores the novel item with several touches itself.

This bizarre animal begs the questions how and why. Of all of the predators on the planet, the star-nosed mole is, inch for inch, pound for pound, the most vicious, voracious, and velocious eater of all. If you think a cheetah is fast to the kill, this little creature can touch, identify, and eat a prey item in 120 milliseconds. There are online video examples. It turns out that, if this mole was even a half-second slower at this predatory behavior, it could not consume enough food to keep up with its rapid metabolism. The acuity and rapidity with which this behavior has evolved are stunning.

And then there is the poor naked mole-rat, which is neither a mole nor a rat. Naked mole-rats are more closely related to porcupines, chinchillas, and guinea pigs than to any other mammals. Recently they were placed into their own family—Heterocephalidae, which means they aren't even a mole-rat. And if that wasn't enough, they aren't really naked, because they have many wiry like tactile "hairs" protruding from their bodies that look like a twelve-year-old boy trying to grow a mustache. But there is a heterocephalidunculus. In 2002, Kenneth Catania and Fiona Remple presented to the world the naked mole-rat homunculus, or heterocephalidunculus. This image was drawn by the modern-day Mrs. Cantlie, Lana Finch, and is a nightmarish picture that demonstrates this species' way of life. Naked mole-rats have tiny eyes that were once thought to be nonfunctional, but a 2010 study revealed that they can indeed use their tiny eyes to detect light. Their hearing is acute, and they can hear the slightest rumbling of an insect in the tunnels they so adeptly construct.

Naked mole-rats are also social animals with a unique hierarchy among mammals. Each colony has a queen and a couple of reproductive males. The rest of the colony are in the worker social caste, and within it individuals assume a place in the hierarchy. One way to tell the queen from the others, besides producing pups every two months or so, is that she always gets to pass other members of her colony on the top in a tunnel. Nearly one-third of the somatosensory apparatus of naked mole-rate is dedicated to its teeth, which it uses to do almost everything except fornicate (and there is evidence that it uses its huge buckies in social interactions, and so perhaps even its large teeth are involved in sex).

What is most interesting about the naked mole-rat is its brain structure. It appears that the cortex of this species has been completely remodeled such that nearly all of the area of the cortex usually dedicated to vision has been converted to the tactile capacity of the teeth. This remodeling is a beautiful example of the plasticity of the mammalian brain, but little is known about how it works. Fortunately, how remodeling of the brain occurs in the context of the senses can be studied closely by use of the mouseunculus.

The normal mouse homunculus (mouseunculus) has the typical larger head and grossly exaggerated snout of homunculus mammals that rely on tactile mechanisms near their head for interpreting the natural world (see fig. 3.1). Rodents typically use the whiskers on their snouts to sense their immediate outer world. These whiskers are long sensory apparatuses that are continually being whisked back and forth and up and down to touch things. These long whiskers are attached to the snout by means of a set of sensors at the base of the hairs. The information from the sensors is transmitted to the neocortex of the mouse, and the information gathered and interpreted in what are called barrels in the somatosensory cortex. Dennis O'Leary and his colleagues at the Salk Institute asked how the touch sensory real estate in the brain is remodeled after shrinking regions of the brain dedicated to this sense.

They posited two possible outcomes when a brain region responsible for assembling touch sense information is reduced in size. First, the organization of the barrels, and hence the means by which the outside information is processed in the brain, could simply be miniaturized. Second, the barrel size and orientation could be altered to a different organization. The easiest way to approach this question is to breed mice with regions of the brain that are smaller than others, and specifically to breed mice with reduced somatosensory cortexes. There is a mutant in mice called *small eye,* which does result in a smaller somatosensory cortex. The mutation is caused by a lesion in a protein called PAX6 that is considered a master switch gene in eye development (a master switch gene is one that needs to be present for any downstream development to occur). Another function of PAX6 is that it controls growth, via gene regulation, in the brain. Unfortunately, *small eye* is also lethal in the embryonic stage, and simply breeding *small eye* mice wouldn't work because they would die before they were useful in the experiment. O'Leary and his colleagues figured out a way to localize the PAX6 gene expression with the lesion to the somatosensory cortex, a spectacular feat in and of itself. With these localized mutants, they were able to obtain mouseunculus maps of mice with smaller somatosensory cortexes. From the mouseunculus so constructed, it is obvious that the rewiring of the brain involves not miniaturization but rather drastic remodeling of the existing plan. Add to this basic but elegant experiment an even more fine-tuned analysis, as explained in box 3.1, and one is led to the conclusion that the developmental trajectory of vertebrates hones the overall neural wiring of the brain. Altering the developmental pathways will drastically influence the sensory capacity of vertebrates.

Such studies shed light on how embryo development affects the limits of sensory perception. Development of the brain is an important factor in how our brains receive and process information from the outer world. Developmental studies can tell us where in the brain certain information is processed. But developmental biology can also tell us about how the organs that sense the outside world stimuli are structured.

We can barely see our noses. Try it. With both eyes open, you can vaguely see some of your schnozzola. Close one eye, and it is a bit better. What we can see more clearly is what is at the end of our noses. Horses in general simply cannot see the tips of their noses, or more appropriately, they have a conspicuous blind spot at the end of their snouts. Although we can see, just barely, the tips of our noses, we also have blind

## BOX 3.1 | BARRELS, BARRELETTES, AND BARRELOIDS

In the mouseunculus, some barrels are missing and others are reduced in size because of the localized PAX6 mutants (see text). The mouseunculi of these PAX6 mutant mice are very different from that of a normal mouse. The PAX6-deficient mouse also has other parts of its brain altered. In addition, the somatosensory cortex (the part of the brain that processes sensory input from other parts of the body), the hindbrain (the back of the brain), and the thalamus (part of the limbic system) are all also affected. It turns out that both the hindbrain and the thalamus have arrangements of patches of neural tissue similar to the barrels in the somatosensory cortex. These patches are even named similarly: in the hindbrain, as they are called barrelettes, and in the thalamus, they are called barreloids. How the barrels develop influences how barrelettes and barreloids develop and, in turn, how the thalamus and hindbrain develop. This result means that there is a hierarchy of signals occurring at the developmental level that fine-tune the ultimate neuronal structure in the brain. Altering the genes in the hierarchy, or how those genes are expressed, can have a huge impact on the sensory capacities of mammals.

spots in our vision. For example, stare at the dot and the X below. Now cover your left eye and get your face very close to the page so that you can see both the dot and the X. Focus on the dot on the left and slowly move your head away from the page. The X on the right should disappear somewhere as you move your head farther from the page. For me, the distance from the page where this happens is about one foot or so from the page. This is the classical definition of a blind spot.

•                                                                          X

Other vertebrates have similar blind spots, but our tiny blind spot is nothing compared to the blind area behind our heads. This huge blind spot we humans have serves as a reminder that we aren't anywhere near the best in our visual proclivities compared to other organisms on this planet. As a species we have a well-defined but somewhat narrow capacity at sensing the outside world. For instance, although we are considered a very visual species, relying on vision for much of the information we need to exist, our range of visual perception is quite narrow compared to other organisms that sense light waves. Consider, for example, field of vision, an aspect of the sense that is a strength. Field of vision is defined as the amount of area that one can see by holding the head stationary. It is influenced mostly by the position of the eyes on the head as well as by how much the eyeballs can range within their sockets. Field of vision can be described as covering two directions: left/right fields and up/down fields. A subaspect of field of vision is how much of the total field is binocular, or stereo. Seeing stereo means that you are seeing an area with both eyes, allowing you to perceive the depth of what you are seeing. Humans have a pretty paltry 180 degrees or so of field of vision, and about half of that is in stereo, or binocular, vision.

As the position of the eyes is turned outward on the head, less of the field of vision of one eye overlaps with the other eye. So, in gaining peripheral vision, an organism gives up some binocular vision. Consider, for example, the pigeon, which has an almost 360-degree field of vision, but only a fraction (say, 30 degrees) of that in stereo. Dogs have a pretty good overall field of vision, nearly twice that of a human, but like a pigeon, a dog's binocular vision is paltry in that it is about half of ours. It should be obvious that almost every variation of field of vision may have been tinkered with during the evolution of animal vision.

But what drives this fascinating characteristic of animal vision? Natural selection obviously drives some of it. Macaque monkeys, for instance, have a very limited field of vision. They can see a little more than 180 degrees. Their binocular vision is a little more than half of this, making for an animal that is not terribly adept at seeing a broad field of vision but very skillful at viewing the world in stereo. Why would a primate need to place a lot of its stock in stereovision? The answer to this question is in where the ancestral primate lived. The ancestral primate was most certainly arboreal, living in trees and adeptly navigating movement in that environment. Try the following, but be sure to have a net or a friend watching you. Close one eye and try to climb a tree (or a ladder). You'll find it quite difficult to judge where to place your hands as you are deciding on your next move. If you feel yourself falling, please open both eyes and adjust. This little exercise demonstrates that stereovision gives us depth perception and makes it easier for us to interpret the physical structures we encounter. Other animals with acute stereovision are almost always those that need to navigate the three-dimensional world by relying on acute depth perception for hunting prey and avoiding predators. Organisms such as pigeons that have evolved to navigate a scary and deadly world of predators don't need much stereovision. They simply need to know that something is sneaking up on them or streaking down toward them. So, natural selection plays a huge role in this phenomenon. But it is not the only part of the story.

Let's consider the octopus (a mollusk), which has a phenomenal 360-degree field of vision *and* no blind spot. Why is there no blind spot in this animal? The answer comes from understanding the evolution of the eyes in both the human and octopus lineages. Eyes have evolved independently about twenty-five times in the history of animal life on this planet. This means that there have been twenty-five independent instances of light-sensing organs evolving in the more than half a billion years or so that animals arose on Earth. The eyes of vertebrates and of octopuses have evolved different structural quirks.

The light-sensing part of the human eye consists of a collection of cells (in vertebrates called rods and cones) that make up the retina of the eye. The retina is connected to the brain through a nerve bundle that in vertebrates passes in front of the retina. Although the part of the retina where these nerves traverse is minuscule, the nerves nevertheless obscure some of the visual field and hence produce the blind spot that occurs in all vertebrates. But in octopuses the eye evolved so that the nerves leading from the retina to the brain are attached to the backside of the retina and do not obscure light hitting the retina. Octopuses therefore have no blind spot. It would be excessively difficult for mammals with blind spots to evolve to overcome their blind spots. And because of the way that mollusk eyes evolved, they were destined to avoid that blind spot. In essence, the structural dynamics of how the vertebrate eye and the octopus eye develop constrain whether a blind spot will exist. Natural selection probably had little to do with the lack of a blind spot in the octopus, although the octopus may currently exploit not having a blind spot.

Scenarios like octopus's lack of an eye blind spot are reminders of three important aspects of evolution when considering adaptations in nature. The first concerns what Richard C. Lewontin and Stephen Jay Gould called spandrels, after the architectural masterwork Saint Mark's Basilica in Venice. After the cathedral was built, artists covered the inside surfaces of the awe-inspiring domes with scenes from the Bible. The artwork fits perfectly on the spandrels. One spandrel features a man pouring water from a large flask into the tapered end of the space at the bottom. It is easy to look at these painted surfaces and think that the spandrels were created explicitly to display art. Not a bad hypothesis, but incorrect, because the spandrels are there to hold up the massive domes of the cathedral, and the artwork, though it "adapts" to the space between spandrels and seems perfect, is an afterthought. So, too, the octopus's lack of a blind spot, though it probably serves an adaptive response today, is an afterthought, driven instead by the architectural wiring of the eye's nerve connections.

A second aspect of evolutionary forces us to recall that we have to resist interpreting everything we see in the natural world as an adaptation, when indeed many are not. In fact, sometimes the solutions to one challenge are compromises to solutions for other problems. On this theme of compromise, consider the eyes of arthropods.

Arthropods are a huge group of animals that include insects, and they have eyes that are called compound eyes. Compound eyes are an ancient structure, as evidenced by arthropod trilobite fossils with exquisitely preserved faceted eyes that are hundreds of millions of years old. Antonie van Leeuwenhoek, noted more for his observation of plaque from his teeth and the motility of his sperm through his famous microscope, was the first to describe the amazing complexity of an insect's compound eye. His small microscope was a handheld device that was backlit by a candle to force light through the sample in the observation chamber. When he mounted an insect eye cornea on the device, he was stunned by what he saw. By moving the candle in back around a little, he was able to arrange it so that it caught the tissue at an angle such that he observed "the inverted images of the burning flame: not one image, but some hundred images. As small as they were, I could see them all moving." He was seeing the light of the candle through the hundreds of little facets, called ommatidia, that make up the compound eye of an insect. Remarkably, each one of the ommatidia of the compound eye is wired to the insect's brain. In addition, with more ommatidia, the more lenses there are and the smaller they get. Diffraction of light eventually becomes a problem, causing blurry focus or poor acuity.

The number of ommatidia among insects varies. Insects with tiny eyes have fewer ommatidia—for example, worker ants that rely mostly on smell to communicate have as few as six. Insects that rely heavily on detecting motion to hunt prey, such as dragonflies, have more than twenty-five thousand ommatidia. Compound eyes are very good at sensing motion because they can detect temporal changes at about two hundred images per second (the human limit is thirty per second, when everything starts to blur). The compromise here should be obvious— minuscule amounts of motion can be detected by the compound eye with lots of facets, but acuity is reduced in the process. Apparently, depending on the need for detecting motion versus acuity, insects have evolved specific numbers of ommatidia as a compromise to compensate one for the other.

The third evolutionary quirk involves how organisms develop. In some cases, the way that organisms develop constrains how morphologies are eventually formed in development. Because of the constraints of development, certain morphologies, even though they might be considered optimal, simply will not be possible to evolve. The placement of our eyes on our heads is constrained by the way the eye develops in vertebrates. The developmental control over the positioning of the eyes is more than likely involved in how the width of fields of vision in vertebrate organisms evolved. This very same developmental control is responsible for the constrained way that our eyes are placed on our heads and how the eyes of other vertebrates find their positions on the head during embryonic development.

In the early days of genetics, it was assumed that one gene was correlated to one enzyme. In fact, George Beadle and Edward Tatum were awarded the Nobel Prize in 1958 for this intriguing hypothesis, which probably rings true for simple single-celled organisms like bacteria. However, for more complex organisms, the story is quite different. The modern unraveling of the real nature of how genes control complex phenotypes probably started in Allan Wilson's lab at the University of California at Berkeley in the 1970s. Wilson and his colleagues recognized that, although humans and chimps are incredibly different morphologically and behaviorally, their proteins are incredibly similar. What this meant to Wilson and his colleagues was that simple changes in the structure of proteins were not responsible for the broad morphological and behavioral difference between organisms. Instead, they hypothesized that changes in gene regulation were more important in producing phenotypic change in evolution than simple point mutations. Consider, for example, eye placement on the face of organisms and hence control over the field of vision in vertebrates.

One of the more important discoveries in biology in the past few decades was the discovery of how gene regulation works to pattern the vertebrate body. And in a way, this phenomenon is also about sensing, since the cells in a developing skull need to recognize where they are, over the developing field of other cells. The sensing is done in pretty much the same way that single-celled organisms sense their outer world, and that is through molecules that can signal the cell and give it a sense of where it is, which in turn is involved in telling the cell what to do. Signaling like this is similar to quorum sensing, only much more complicated, but the general idea of quorum sensing is there. Signaling molecules work by binding to other molecules in the cell. For some signaling systems that require the precise development of structures in a vertebrate body, the amount of signaling molecules present near a cell will dictate what the cell does. This is because signaling molecules work by gradient diffusion. In general, genes in cells have different concentrations of specific signaling molecules that they need to be turned on or regulated to start producing proteins. If there is variation in the concentration that gets a gene pumping out product, then a gradient of the signaling molecule will induce different outcomes at one end of the gradient (say, the low-concentration end of the gradient) as compared to the other end of the gradient (the high-concentration end).

This scenario is basically how the position of the eyes on the head is determined in vertebrate heads. The signaling molecule in question was first discovered in *Drosophila melanogaster* (the fruit fly) and was subsequently found in the genomes of vertebrates. Genes that produce proteins and interact with this signaling molecule pathway got named after hedgehogs. The embryos produced by mutants in hedgehog lesions result in stubby, hairy little creatures that don't live past early developmental stages. In an orgy of silly gene naming—and *Drosophila* biologists are perhaps the worst of all silly gene namers—one of the important signaling molecules was named Sonic Hedgehog (Shh) after the video game and cartoon character. Other hedgehog genes like Indian hedgehog and desert hedgehog and even tiggywinkle hedgehog (see Bea-

## BOX 3.2 | HOW TO MAKE A CYCLOPS

The signaling molecule Sonic Hedgehog (Shh) creates a gradient in the developing vertebrate embryo that controls a series of genes that in turn determine cell type in the developing brain and skull. The figure shows the gradient as being at the very anterior part of the face. The light grayish strips in the diagrams show where Shh is expressed. In the far-left panel, this signaling protein is turned on full blast in the normal developing embryo. It signals the production of all of the proteins and all are made (white, light gray, dark gray, and black), and the normal eye field knows where to develop above the black protein. The gray protein nearest to Shh needs the most Shh in order to be expressed, and the different-shaded gray, white, and black proteins need intermediate amounts. In the second panel, some of the Shh has been stripped away. When this happens, the light gray protein closest to the Shh protein doesn't get expressed, as shown in the third panel. As more and more of the Shh gets stripped away, the white, light gray, and black genes don't have enough Shh to turn on, and so their activity is ablated, as shown in the fourth panel. The fifth panel shows the result when all of the Shh gradient is removed, and the sixth panel shows that the eye field has moved to the very bottom of the developing brain and that both eyes have overlapped, producing a Cyclops-looking creature.

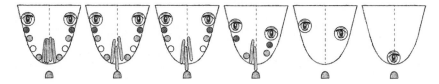

Figure 3.2. Thomas Jessel's hedgehog gradient explanation for where eyes develop on the head. The different filled dots represent four proteins that need to be made for the eyes to be placed normally on the head. The gray strips near the bottom of the developing brain represent the expression amount of Shh, which controls the production of the four proteins. Where the eyes are pictured represents their final position after development. See box 3.2 for details.

trix Potter) were also coined. But we will concern ourselves with Shh here. An elegant way of explaining this difficult series of events that I use here has been presented previously by Thomas Jessel and is shown in box 3.2 and fig. 3.2.

This Cyclops phenomenon actually occurs in nature. Cattle or sheep that eat the false hellebore (plants in the genus *Veratrum*) ingest large amounts of alkaloids that are sequestered in the plants. It turns out that these alkaloids block production of proteins in the hedgehog pathway, hence producing a situation like the one in the far-right panel of the figure. The Cyclops produced by the knockdown of Shh production in these animals is striking, but it also gives us a way to visualize how the various placement of eyes on the heads of vertebrates might have evolved. Many genes are involved in the layout of the nervous system in the head and eyes and thus influence where the eyes are placed. But tweaking the signals that these genes are interacting with slightly is a perfectly logical and productive way to think about how nature can tinker with field of vision. Human development has settled on a specific field of vision connected sharply to how the nervous system and eyes evolved, and we are stuck with the relatively paltry field of vision we have. So, we are not very good with respect to field of vision compared to other animals, but at least we know why.

# 4 A MATTER OF TASTE (AND ODORANT) RECEPTORS

*Smell and Taste Reception in Animals*

*"In the land of the skunks, he who has half a nose is king."*
—Chris Farley, comedian

Most animals have evolved to be very discriminating about what they eat. If we smell or taste something incredibly rotten, for example, we avoid eating the rest of it. This response more than likely evolved as a means to deal quickly with the "I eat that" binning process I discussed in Chapter 2. All of our senses probably help us make these quick "I eat that" decisions, but how we do it has a deep evolutionary history. Remember that vertebrate brains have three basic layers of organization. The innermost, most primitive layer probably has its roots in our ancestral connection to early vertebrates, and it holds the brain stem and the cerebellum. The next layer, consisting mostly of the limbic system, adds complexity to how such information as smell and taste is interpreted. And the final layer, the cortex, adds to an even more refined way of interpreting the information from our senses.

How our sense of taste is interpreted in this layered brain is a great example of how these three layers are integrated. Taste interacts with our reward system through what neuroanatomists call the cortico-basal ganglia-thalamic loop. This loop is a set of neural tissue pathways that traverse the three major brain regions implied by its name. The most prominent reward systems in vertebrate brains are the gamma-aminobutyric acid (GABA) and dopamine wired neurons. GABA and dopamine are two small molecules that make their way to our brains and interact with receptors embedded in the membrane of neurons to trigger action potentials in the brain. The dopamine neurons in particular have a huge role in the evolution of how animals use the reward system.

Pleasure is a big part of training organisms to repeat things that are beneficial to them. It makes sense that if something is both beneficial and pleasurable, an organism will seek more of whatever triggered this reaction, such as sex, something that tastes good, or, sadly for the long run, a pleasure-inducing drug. When we taste something rotten, our brain dopamine levels fall drastically, telling us that we don't want to taste any more of the bad stuff. But if we taste something nice and sweet or very nutritious, dopamine levels in the brain increase, and the reward system responds by saying, "I want more." The dopamine tells us to ingest as much as we can, but because it is temporary, at a certain point we are satisfied and stop ingesting the item. Drugs such as cocaine and heroin exploit this system and hijack the brain. Instead of causing a transient dopamine concentration, the molecule establishes itself at a high plateau level, producing a craving for more of the drug on a scale that goes on a runaway course and results in addiction.

How this system starts is very similar to how odors are processed—it begins with small chemicals or molecules and a chemoreceptive sensation. Tastes emanate from the combinations of small molecules that we ingest with food or in the air or in beverages and are processed by their interaction with taste receptors in the mouth. That information is then transmitted to the brain, where the information is interpreted. The repertoire of taste receptors consists of five basic kinds: bitter, sweet, umami, sour, and salty. Carbonation and fattiness also probably qualify. Taste receptors occur predominantly on the tongue but have also been found elsewhere, for example, in the tissues of the airway and in the small intestines.

Researchers have characterized several kinds of taste receptors that influence sweet, umami, and bitter taste reception, called TASs (named so for the first three letters in the word "taste"), that act a lot like the odorant receptors for smell. TAS1s are involved in sweet and umami reception, and TAS2s are involved in bitter taste reception. Receptors for salty and sour have also been proposed, but less is known about them. There is one major candidate for sensing salty, and it is a gene that is also, oddly enough, involved in polycystic kidney disease (PKD) called PKD2L1 (2L1 indicates the kind of PKD gene involved). It is an ion channel receptor that is also involved in the sensing of acid. Ion channel receptors are proteins that reside in the membrane of the nerve cell and are responsible for moving specific kinds of ions across the cell membrane. In so doing, these ion channels start the taste response by initiating action potential as the ions move across the membrane. Other receptor molecules for both sour and salt are sodium channel (SC) receptors, called SCNNs, of which three (SCNN1a, SCNN1b, and SCNN1g) are thought to be the major conveyors of information about salt and sour to the brain. Suffice it to say that there may be more sour and salty receptors out there. Like odorant receptors, the number of genes for these signaling molecules that are found in the genomes of animals is an interesting phenomenon. Humans have about seventy of the sweet, umami, and bitter receptor genes (the TAS1s and TAS2s), and a few genes are found in the sour and salt categories. The animal world, however, is much more interesting when it comes to taste (fig. 4.1).

In all likelihood, animals are not discerning enough to place tastes into five categories and simply place them in three: Yum! ("I like that"), Yikes! ("I don't like that"), and Ho-Hum ("I don't care"). Those of you who are cat lovers can try to bribe your feline friends with a sweet. Try it, and you will find that your cat doesn't give a meow about sweets. This lackadaisical attitude of felines toward sweets is not a part of their cool cat demeanor but rather the result of the fact that cats simply can't taste sweets well. Sweet taste reception is implemented by two of those seventy or so TASs, called TAS1R2 and TAS1R3. In animals that can taste sweet (humans included), these two TASs form a coupled protein. When something sweet enters the mouth, the sugary compounds from the sweet item bind to the coupled protein, and this produces a signal that goes directly to the brain and to the cortico-basal ganglia-thalamic loop, where it is interpreted as "I like that" very likely because sweet things have lots of important carbohydrates in them. Cats have a long deletion in the TAS1R2 gene that knocks out the function of this receptor protein, causing it to be classified as a pseudogene (a gene that is present in

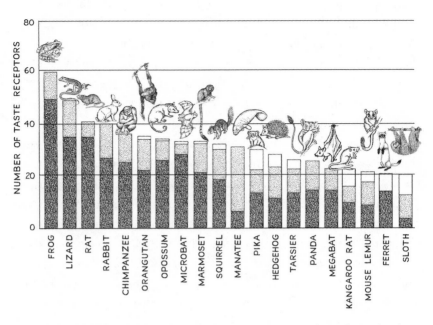

Figure 4.1. Bar diagram showing the number of intact, pseudo, and truncated taste receptors in a broad range of vertebrates.

the genome but does not make a proper protein). This simple genomic change kills most of the cats' chance for detecting sweet tastes, although some researchers think that cats may be able to detect very large doses of sugar as sweet. Researchers know that the loss of this long stretch of the TAS1R2 gene occurred in the felines' common ancestor, which means that the big cats, such as lions and tigers, along with domestic cats, cannot taste sweet. It turns out that canids, the closest relatives of felines in the order Carnivora, can taste sweets. Even the panda, that charismatic bamboo-loving bear, has intact sweet receptor genes. Pandas can taste sweet and prefer sugary water to tasteless water when offered such libation. And this probably explains why we hang our backpacks with sweet granola bars in them out of the reach of any species of bear when we hike in areas where these sweet-toothed carnivores live.

An interesting aspect of the evolution of the ability to taste sweet in carnivores arises when thinking about the distribution of sweet taste ability. Is the loss of sweet taste in felines the cause or the effect of felines' preference for meat? It is difficult to determine how the loss of function (also called pseudogenization) of TAS1R2 is involved, but the

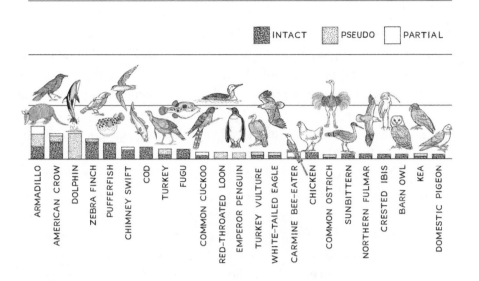

phenomenon does remind us that to jump to the conclusion that the loss of the function of the gene is adaptive for meat eating is erroneous. It might be true, but equally likely, the lost region of the TAS1R2 gene occurred through a chance event in the genome of the common ancestor of the felines and was later coopted to reinforce felines' meat-eating characteristics. This alternate scenario is a good example of what Stephen Jay Gould and Elizabeth Vrba called an exaptation. The trait evolves for some fairly mundane reason and is later exapted or coopted by more visible trait systems (in this case, felines' preference for meat diets).

If both smell and taste are chemosensory, then how do smelling and tasting differ? Insects are a great example to examine the differences. Besides the fact that in insects the two senses use different receptors, one can always distinguish them by the following logic: Smell is implemented by the detection of gaseous molecules—usually on the antennae. Taste is accomplished by direct bodily contact with the item being tasted. So, what body parts do the tasting in insects? Insects such as flies have mouths—almost. They have mouthparts, and that is a way of saying it's pretty ugly in there. The mouths of those alien hunters in the *Predator* movies are modeled after insect mouthparts, and those aren't pretty at all. The fly mouthparts have taste receptors on them called gustatory receptors, or GRNs (the N stands for the variant)—the fly has about seventy GRNs—but they have no relatedness to the taste receptors found in vertebrates. Clearly, though, flies can detect bitter and sweet as well as water and carbonation. It is not surprising that flies and other insects have gustatory receptors embedded in those ugly mouthparts that come into constant contact with food. And it really shouldn't surprise us that GRNs are also placed on the wings and legs, and even on the egg-laying apparatus of females, because insects use these appendages to sense the nutrients in the surfaces they touch.

The range of number of GRNs in the genomes of insects is impressive—from eight in lice (*Pediculus*) to more than two hundred in the flour beetle (*Tribolium*). The number of odorant receptor genes is also weakly positively correlated with gustatory receptor genes in the genomes of insects, suggesting that the rich get richer with respect to such receptors and with respect to precision of those two senses. Insects that rely on one or a few sources of food would be expected to have fewer taste receptor genes. They need to know whether they are eating what they should. But insects that are a bit more adventurous with their diets (also known as polyphagous insects) might be expected to be more discerning with respect to taste. This is indeed the case with some polyphagous insects, where their genomes contain a couple hundred gustatory receptor genes. But the overall correlation is weak, and a lot more research is needed on how insects taste and how that relates to the evolution of what they eat.

The range of taste receptor genes in vertebrates is equally broad as in insects. In animals, the bitter taste receptors are perhaps the most interesting, and this makes good sense with respect to placing things into the "I eat this" bin strategy. Bitter taste is probably the most important in discerning what an animal should avoid. Sweet-tasting items are a fairly easy choice, because an organism will want to consume the carbohydrates in them. But an organism must be much more discriminating with bitter foods and can't simply avoid them all. This may be why bitter taste receptors (TAS2s) have a large range of copy number variation in vertebrates. Because herbivores get far less nutritional value for their diet items (plants in general are less rich in calories and other nutritional components than meat), they can ill afford to reject a plant that might have nutritional value. Diyan Li and Jianzhi Zhang argue that herbivores need to be more picky about the plants they encounter. The increase in resolution of bitterness assists them in being picky about rejecting bitter things while still being able to avoid nasty-tasting foods (see box 4.1).

Cats also appear to have quite a few truncated bitter taste genes and a lot of pseudogenes, but they still retain some genes for bitter taste. This exercise of counting genes led to the fascinating discovery that marine mammals (specifically dolphins—see fig. 4.1) have no functional genes for bitter taste. In fact, all cetaceans appear to have experienced a massive loss of bitter and sweet taste genes, according to findings by Ping Feng and colleagues, who examined twelve cetacean species for the bitter and sweet receptor genes. The loss of these receptor genes means that cetaceans have lost four of their five kinds of taste—sweet, umami, sour, and bitter. Cetaceans still retain salt taste, suggesting an evolutionary mechanism for how these marine mammals taste things. However, Feng and colleagues conclude that cetaceans really don't taste, nor do they need to. The high concentration of salt in the marine environment overwhelms most tastes, and many species swallow prey whole, without tasting their food at all: the other four senses have simply been lost through disuse. The marine environment is hard on TAS2 bitter taste

---

## BOX 4.1 | TASTE RECEPTORS

Taste receptors range broadly across the vertebrates. Bitter-taste receptors, like olfactory receptors, include some pseudogenes and some truncated genes. It appears that for some vertebrates a good proportion of the genes in the genome are either pseudogenes or truncated. The number of receptors ranges from three in some birds and fish to up to seventy (only half of which are functional) in the guinea pig, and about sixty (of which more than fifty are functional) in a genus of frog called *Xenopus*. By analyzing the diets of various vertebrates and examining their taste receptors, Diyan Li and Jianzhi Zhang have been able to make predictions about how the number of bitter-taste receptor genes influences diet (or vice versa). Although they warn that interactions of taste receptors with the ecology of an organism are complex, they conclude that herbivores have more TAS2 receptors than omnivores and carnivores. It appears that taste has played a central role in helping vertebrates decide which items to place into the "I eat this" bin.

---

genes. The manatee, the only other mammal in the study that lives in a marine environment, has had 75 percent of its bitter taste genes pseudogenized, rendering them inactive. But even though it has fewer TAS2 genes and fewer taste buds in its mouth, the manatee can still taste. Perhaps its herbivorous diet and the fact that it actually chews its food has prevented the entire gene family from being blanked out.

Birds show a paucity of taste receptor genes as all birds have very few TAS2 receptor genes (fig. 4.1). The American crow and finches have the most, at seven intact bitter taste receptor genes, and penguins have none. Again there is an ecological correlation of diet with the number of TAS2 receptor genes. Using the same reasoning of Li and Zhang, Kai Wang and Huabin Zhao point out that herbivorous (and some insectivorous) birds tend to have more TAS2 receptor genes. The penguins are another story, because they have also been examined for the loss of taste receptor genes for the four other tastes. The results of that study demonstrate that penguins have lost their sweet, bitter, and umami receptors, while they have retained sour and salty putative receptors. Other birds appear to have retained nearly all of these taste receptors, except for the loss of sweet receptors in some bird lineages. Penguins, unlike other birds, don't have taste buds in their tongues, and they swallow their food whole, obviating the need for taste other than salty. It should be noted that these inferences are possible only because of the increased capacity to sequence whole genomes of microbes, animals, and plants. And as more and more organisms have their genomes sequenced, the prospect of uniting feeding ecology with the genetic and molecular aspects of taste will be realized.

Anyone who has been in New York City in mid-July might argue that our sense of smell is pretty good at detecting unpleasant odors. The smell is so bad in July and August in New York City that it prompted a children's book author to call it Phew York City. One species of insect commonly seen flittering over the garbage that makes the terrible smells in Phew York City, called *Drosophila melanogaster,* has been the workhorse of biology for more than a century and has contributed greatly to our understanding of smell. Charles Woodworth first suggested in the early twentieth century that this tiny fly, also known as a fruit fly, would be a good experimental animal. Because it reproduces rapidly (every ten days) and is easy to grow in the lab (a little banana and apple sauce mixed with oatmeal and vinegar usually does the trick), it was touted as the ideal lab animal. The geneticist Thomas Hunt Morgan settled on it as his experimental organism of choice in the early 1900s. He was quickly rewarded for choosing this tiny fly by discovering several very visible spontaneous mutants (white eyes and curly wings, among many) that he could use to work out the rules of crossing over of genes on chromosomes. But the tiny fruit fly also has played an amazing role in understanding the mechanics of olfaction. William Morton Barrows recognized that the fly was more than likely using odors to mediate its behavior in 1907 with this statement: "The fact that the fermenting fruit upon which they feed is continually generating alcohols, and other related compounds, led me to suspect that it was these substances that served to attract the flies."

Barrows devised ingenious experiments to pin down that the flies were responding to chemical odors. It wasn't until almost fifty years later that more refined methods were used to generate *Drosophila* olfactory mutants. Several inventive devices were constructed to test mutant flies in the lab for alterations in their olfactory capacity. The most common is the "Y-shaped tube," in which the "Y" is placed upright. The air is sucked out of the "Y" to remove any lingering odors. The odorant under study is placed at the end of one of the slanted parts of the "Y," and the other is left unscented as a control. Flies like to climb (the technical term is that they are "negatively geotactic"), so they will climb up to the junction of the two slanted parts of the "Y." Once there they make a decision based on whether they like or dislike—or literally can't stand— the odorant. Very clever ways of counting and interpreting the data have been developed, and they lead to identification of flies with mutations that either lose or gain capacity to detect specific odorants (box 4.2).

Using rats, Linda Buck and Richard Axel looked at odorant phenomena in vertebrates. Their landmark study revealed a large and diverse array of odorant receptor genes in mammals. And in the end Buck and Axel realized that the odorant molecules indeed interact with the odorant receptors like locks and keys. If the odorant receptor has the right "lock" for the odorant key to fit in, then the receptor will induce further reactions in the cell that lead to neural transmission to the brain that the odorant is there.

It didn't take long for *Drosophila* biologists to jump on the Buck-Axel

## BOX 4.2 | FINDING ODORANT MUTANT FLIES

A simple metric is used to determine whether a fly has an odorant mutation. The metric is based on the ratio of odorant-sensitive flies to control flies (who aren't sensitive to the odor). As described for the Y-shaped tube experiments in the text, flies are given a choice to react to specific odors and then counted on the basis of their reaction. If there is a departure from random movement of the flies (50 percent to the control side and 50 percent to the odorant side), the mutant is kept and analyzed further for its odorant capacities. John Carlson and his colleagues at Yale University built a very clever apparatus to detect the odorant sensitivities of flies in 1989 using discarded lab items such as small test tubes and pipette tips. Next, they immobilized the mutants and examined the fly organ that mediates the odorant effects to the fly brain—the antennae. In this way, they isolated six odorant mutants and were able to correlate these with alterations in the function of the antennae. It could be that every time one walks into a department store cosmetics department someone is doing the same experiment with humans, but I doubt it. The point is that the discovery of a slew of mutants in this way led to the genetic description of olfactory mutants in the tiny fruit fly, which in turn led to a more complete understanding of how smell works. Specifically, researchers reasoned that since odorants were molecules, receptors were detecting them. The mutants discovered by *Drosophila* workers were seminal in coming to this realization.

bandwagon, but they had an advantage—the sequence of the genome of their favorite organism was nearing completion a year or two ahead of the human genome, so they were able to get a complete view of the repertoire of *Drosophila* odorant receptors and how they worked. They found at least sixty-one odorant receptor genes in the *D. melanogaster* genome. None show enough similarity to vertebrate odorant genes to warrant calling any the same as vertebrate receptors. In fact, the *D. melanogaster* odorant receptor genes show extreme sequence divergence with other insect genes, indicating that these receptors change rapidly over evolutionary time. The structure of the proteins coded for by these genes is very interesting, in that all follow a general theme of being embedded in the membrane of the organ that receives odors—in the case of insects, the antenna. The typical odorant receptor protein is threaded through the membrane of the odorant receiving organ's cells with what are called transmembrane domains. The part of the odorant receptor sticking out of the cell will specifically bind compounds that then trigger intracellular reactions in the receptor cells, which then signal the brain that the specific odorant is there. *Drosophila*'s sixty-one odorant receptor genes are paltry compared with the more than one thousand found in the nematode, and even more paltry compared with the approximately nineteen hundred in the elephant. But counting odorant receptor genes as a measure of the smelling ability of an organism is complicated.

The first complication is that not all genes in an organism's genome are expressed. So, although one might be able to detect the presence of sequences that are commonly found in a particular kind of gene, that doesn't mean the gene is active. When this occurs, the genes are called pseudogenes, as I noted earlier. The lack of expression of pseudogenes usually is caused by the occurrence of a stop codon in the gene, leading to a truncated and nonfunctional gene. The stop codon is a signal to the protein translation machinery of the cell to end translating a gene into protein. The range in number of odorant receptor genes in vertebrates is impressive (see fig. 4.2 and box 4.3).

The second complication is that even the sixty-one odorant receptor proteins of *Drosophila* can still accomplish a lot of smelling. Many odors can be discerned quite exquisitely by a small number of receptors because of the combinatorial nature of how odorant processing occurs in the brains of animals. A single odorant can bind more than a single receptor, and hence a neuron can have multiple responses when enervated by multiple receptors. In addition, information from neurons with receptors converges to focal processing points that are called glomeruli. In this structure, multiple signals can be combined to be more discerning about odors. The combinatorial nature of odor reception means that as the number of receptors increases, the ability to sense odors also increases, but not linearly. Rather, the increase in potential smells surges exponentially as the number of genes increases.

In addition to odorant receptors, the nasal passages of vertebrates hold another kind of molecule that works in combination with the odorant receptors. These proteins are called TAARs (trace amine-associated receptors), and they function just like their names indicate—they detect trace amounts of small amine molecules. To increase their capacity to

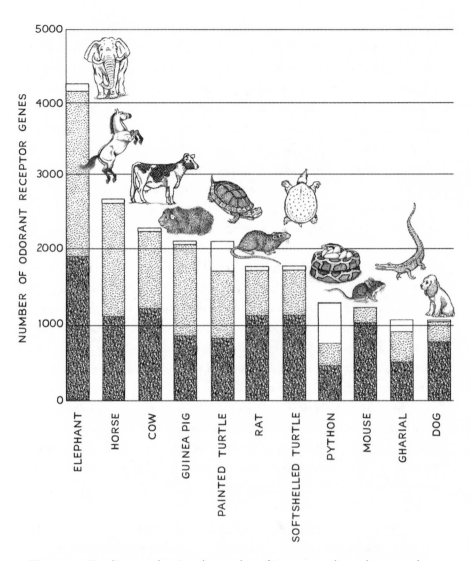

Figure 4.2. Bar diagram showing the number of intact, pseudo, and truncated odorant receptor genes in a broad range of vertebrates.

sense odorants in vertebrates, these receptors are found in different combinations on the olfactory organ. The response to these combinatorial messages sent to the brain produces behaviors that are essential to the organism's survival. Linda Buck and colleagues have shown that, while some of the TAAR/odorant receptor responses in the brain are innate and aversive, such responses can be modulated by other odorant receptor signals to the brain. The neat part of this discovery is that the sorting

INTACT   PSEUDO   TRUNCATED

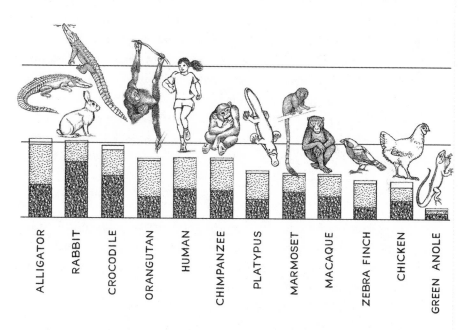

ALLIGATOR   RABBIT   CROCODILE   ORANGUTAN   HUMAN   CHIMPANZEE   PLATYPUS   MARMOSET   MACAQUE   ZEBRA FINCH   CHICKEN   GREEN ANOLE

out of the odorants is being done in the brain based on information from a large amount of stimulation of the nose. Our ability to detect trace amounts of a substance is nothing compared to some insects (box 4.4). Perhaps the most famous smelling feat in animals is the capacity to smell and process information from pheromones. Among the insects, Lepidoptera (moths and butterflies) are particularly good at detecting small pheromone molecules at great distances (up to two miles).

Some odorant researchers think that humans are not very good at smelling (see box 4.3) and are a mediocre species with respect to the number of odorant receptor genes in our genomes (we are in the bottom third of organisms examined at the genome level with respect to total number of odorant receptor genes). But we are actually quite typical in

Humans have about eight hundred odorant receptor genes, but only about half of these are active. In almost all vertebrates the number of pseudogenes plus truncated genes exceeds or closely approaches the number of functional genes. Another factor involved in overall gene number is that genes can be gained or lost independently. Chimps and humans had a common ancestor more than six million or so years ago. This common ancestor had a unique combination of odorant genes that it passed on to both the chimp lineage and to the human lineage, but with some changes. For example, in going from the common ancestor of chimps and humans to our lineage, eighteen odorant receptors are gained and eighty-nine are lost. For the chimpanzee lineage, eight genes are gained and ninety-five are lost. The numbers of gains and losses for other taxa to their most recent common ancestors are similar, indicating that as species diverge, their olfactory capacity is fine-tuned by the loss and gain of genes that detect specific odorants. Odorant receptor gain and loss, then, plays a major role in the overall capacity of an organism to smell.

that about half of our odorant receptor genes are pseudogenes. In addition, the convention, since the early twentieth century, has been that humans can detect about ten thousand odorants. In 2014, however, Andreas Keller and colleagues obliterated this notion.

Keller and his colleagues started with 128 known odorants. They then mixed ten, twenty, or thirty of the common ones in jars. For a single experiment (called a discrimination test), they would make three jars—two that were identical and one with something else. The mixtures were paired up in trials such that the odd-man-out jar in some pairs had no odorants in common and others were almost identical. Each volunteer in the experiment was given 260 discrimination tests, and the results were tabulated. All that was needed was to figure out where the ability to detect different odorants in the odorant mixtures drops off. Keller and colleagues assessed this by the amount of overlap the mixtures had. So, they mixed odorants to produce, for instance, 25 percent, 50 percent, 75 percent, or 95 percent overlap in the test jars. If there was 0 percent overlap (the two test jars had no odorants in common), most respondents could easily discriminate between the two jars. If there was 97 per-

## BOX 4.4 | THE LIFE OF A CATERPILLAR

The French entomologist Jean-Henri Fabre was the first to describe the phero-mone response in detail in the late 1800s. He had found the cocoon of a great peacock moth (*Saturnia pyri*) and began a watch over it (fig. 4.3). Soon a beau-tiful female emerged from the cocoon. He placed the female in an enclosure to allow the eclosion process to be completed and went to bed. The next morning, he woke up to find tens of male great peacock moths clinging to the enclosure. He collected the males and left the female in the enclosure overnight again, this time becoming a peeping tom by staying up the night to observe. He continued this ritual for several days and nights, over which he collected 150 or so male great peacock moths. A prolific writer, Fabre was single-handedly responsible for a resurgence of public interest in insects and entomology in the late nineteenth century. The following quotation from his book *The Life of a Caterpillar* describes what he observed of the great peacock female and shows why he was so good at popularizing entomology: "As I said it was a memorable evening, this Great Peacock evening. Coming in from every direction and apprised I know not how, here are forty lovers eager to pay their respects to the marriageable bride born that morning amid the mysteries of my study. For the moment let us disturb the swarm of wooers no further." Although he didn't know it, Fabre was describing the pheromone attraction in insects. He even admitted that he didn't know how the males were "apprised" that a female was nearby. Decades of research on pheromones have deciphered how the males were "apprised," and put bluntly, they simply smelled the presence of the female. Oh, and yes, we vertebrates have and can react to pheromones, too.

Figure 4.3. Fabre's great peacock moth.

cent overlap (one odorant difference in the thirty odorant mixtures) and no subjects could identify the difference, then the drop-off point would be 97%, and so on. The drop-off point turned out to be in the 50 to 60 percent range, meaning that if the odorants overlap by less than, say, 57 percent, most mixtures are distinguishable. Above that percentage of overlap, the odors from the jars are mostly indistinguishable. The techniques involved in interpreting these data involve complex statistical and mathematical processing and are beyond the scope of this book.

But let's look at what the math actually solves in these experiments, and we should be able to understand the repercussions of the study. For the thirty odorant mixtures there are $1.54 \times 10^{29}$ possible combinations, and for ten odorants there are $2.27 \times 10^{14}$. These are very large numbers, and not all combinations will be discerned by the human nose and brain. The trick is to see how many of these can be discriminated by the human odorant receptor apparatus. Doing the math results in the astounding inference that on average humans can discriminate 1.72 trillion different combinations when thirty odorants are combined. That's 1,720,000,000,000 different combinations! *Drosophila* has been estimated to discern 65,000 different odors, and other mammals more than likely have the same odorant discrimination capacity of humans. While controversial this number of potential odorant combinations very significantly outrivals the range of sounds, taste, and sight humans can detect. Even if overestimated by several orders of magnitude the estimate implies that what at first appears as our Achilles' heel sense turns out to be one of our best compared with our other senses.

# 5  ALL EARS (AND EYES)

*Animal Hearing and Sight*

"Bumblebee bat, how do you see at night?—I make a squeaky
sound that bounces back from whatever it hits. I see by hearing."
—Darrin Lunde, entomologist and author

Because hearing and balance involve the inner workings of the ear
and are closely related in how they detect and transmit information to
the brain, it is logical to discuss them together. But talk about a Rube
Goldberg apparatus! The inner ear, where the structures that handle
both of these senses are found, is astonishingly intricate and convoluted,
with many moving parts. If this structure from many different vertebrates were shown to an engineer, there would probably be no way she
or he would figure out what it was for, which is a reminder that evolution does not produce perfect or even logical structures in organisms.
But one feature would be certain: the engineer would recognize that
they all have the same basic parts and so are modifications of one basic
structural apparatus.

If the engineer were then shown insect hearing organs, she or he
would be even more confused, because the hearing apparatus of these
animals more than likely evolved many times independently, yielding
structures that don't seem to have commonality at all even though they

all have the same function. The ancestor of all insects was deaf, and the story of insect ears is about converting an existing structure in the insect body to a hearing organ.

But why would an insect want to hear in the first place? After all, hundreds of thousands of species of insects living today don't have hearing apparatuses and so don't really hear. For instance, of the 350,000 known beetle species, only a small proportion have hearing. It is evident through evolutionary analysis that hearing arose independently in many groups of insects at about sixty-five million years ago. Something big must have triggered this shift in sensing the world in insects. In one word, it was bats.

Bats hunt by a process called echolocation. In fact, some cave-dwelling birds and some marine mammals also use echolocation to "hear" their surroundings. In echolocation, an animal makes a sound and then literally listens for the echo of the sound. By measuring the time the sound takes to bounce back to the ears, the echolocating animal can get a good idea of whether there is something else out there, and if so where it is. Because the animal has two ears, this makes for an efficient sound reception system—the animal can actually hear differences in the two ears and use them to further refine where objects are, including flying insects. So, what an insect with hearing does is intercept the sound from the echolocating bat and use that information to attempt to avoid the predator. Bats and flying insects have engaged in an ever-escalating arms race over the past sixty-five million years with respect to echolocation and ears. Oh, and of course, as with most things in nature, hearing in some insects has been involved in enhancing copulation.

There are two major groups of bats: microbats (Microchiroptera) and megabats (Megachiroptera). The Megachiroptera include the fruit bats and other relatives and, except for one species, do not echolocate to find food. The one species of Megachiroptera that does echolocate uses a short-click vocalization that is very different from the vocalization used in echolocation by microbats.

A microbat echolocates by emitting high-frequency sound waves from its larynx that range between 14,000 and 100,000 hertz and are then broadcast from its mouth or its nose. The more vibrations per second, the higher the frequency. Humans perceive sounds in the range of 20 to 20,000 hertz, so we can hear only some bat echolocation screams.

Microbat vocalizations are also somewhat species specific, because in adapting to a particular habitat a bat will also adapt the echolocation frequency for the environment. Even though some bat species overlap in the frequency of their echolocation calls, researchers have developed echolocation call libraries for bats much like the bird-call libraries used by ornithologists.

But how does an organism develop ears? One way is through the slow but steady accumulation of change by which Charles Darwin thought all life evolved. This process of imperceptible change was articulated in *On the Origin of Species* as a universal principle for how life evolved.

Darwin was smitten by the work of the geologist Charles Lyell, and Lyell's *Principles of Geology* was one of the books Darwin took with him on the voyage of HMS *Beagle*. Lyell made clear that changes in the Earth's geology, such as the emergence of mountain ranges and erosion, were best described as gradual, and Darwin felt likewise, that change in living organisms over evolutionary time was also gradual. This concept of gradualism was taken up by early evolutionary biologists and dominated the way evolutionists thought about change in the organic world until the 1970s, when Niles Eldredge and Stephen Jay Gould suggested that evolution might occur in a punctuated fashion. The second mechanism conjures up the "hopeful monster." This is a term first used by the nineteenth-century zoologist Richard Goldschmidt and resurrected by Gould in the 1970s to explain the appearance of novelties in the divergence process. We have already discussed one potential hopeful monster in Chapter 3—the Cyclops—which, although it is monstrous, might not be so hopeful as a survivor in nature. One problem with hopeful monsters is that, although large-scale changes in traits do occur, they are usually accompanied by lethality in the individuals with the mutation for the change. This side effect, which is very common in experimental genetics and developmental biology, occurs because genes usually interact in more than a single pathway of development. This phenomenon is called pleiotropy and is another evolutionary pathway that organismal life can take to generate novel structures and behaviors. So, although a mutation in a gene might produce a hopeful monster in one pathway, it will be lethal in the other pathway or pathways that it affects. How hearing and balance evolved in insects is a beautiful story of evolution at work in the realm of hopeful monsters and pleiotropy.

The ancestors of hearing insects, and all insects for that matter, probably had pretty good balance. Part of balance in insects is accomplished with what are called chordotonal organs, small patches of neural cells that communicate to the brain how stretchy the tissue between body or appendage segments is. By judging how much two adjacent segments are stretched relative to each other, the insect can determine where its body is in space. This is a basic proprioceptive mechanism that insects need for efficient locomotion. The chordotonal organs are made up of mechanoreceptor cells that also react to vibrations and hence were preadapted or exapted (see Chapter 4) to be able to detect sound of different frequencies. Anywhere a junction of segments of structures exists, a chordotonal organ is usually placed as a proprioception organ to assist in balance. What this means is that the potential for the development of an ear exists in bizarre body parts with this kind of proprioception, such as on the abdomen and legs. And in fact, it appears that evolution has taken advantage of converting as many of these primordial chordotonal organs as possible into ears.

Another possible evolutionary mechanism for ears to evolve is, as I mentioned above, through exaptation. Some insects have exapted the chordotonal organs of their antennae to make ears, and some have also used these organs in their mouthparts. Different groups of insects have accomplished this exaptation independently about ten times. Some insects have the antennal version of the ear. These are mostly diptera, like flies and mosquitoes, where the proprioception organ, called Johnston's organ, has been converted to a hearing organ to interpret wingbeat sounds in members of their species. Wingbeats are an important part of courtship in these dipterans. Other insects have used the chordotonal organ hinges in their wings and legs and converted these to ears. Finally, the area between adjacent segments has chordotonal organs, too, and these have been converted into ears several independent times. In some insects these organs produce ears on the abdomen of the insect called tympanal ears. Most of the time a single pair of chordotonal organs is transformed (one on each side of the insect), but in the bladder grasshopper *Bullacris membracioides*, six chordotonal organs on each side of the insect have been converted into ears, making for twelve abdominally located tympanal ears.

It is more than likely that the vertebrate ears with which we are so familiar arose only once (thus evolution works one way in one group of organisms and another way in other groups of organisms). Vertebrates have been around for more than five hundred million years and insects for about four hundred million years. If the evolutionary process was playing on a level surface in both groups, then we might expect ears to arise independently about the same number of times in the two lineages. But ears arise ten times more frequently in insects than in vertebrates. Understanding why vertebrate ears have stayed so static requires following five bones near the jaw and side of the head through the evolution of vertebrates.

Let's follow the arrangement of these bones in three kinds of organisms (bony fish, reptiles, and mammals) and in so doing learn how traits like the jaw and the ears can change over time. To make the exercise fruitful, we need to consider which structures are ancestral and which have evolved—so we need what evolutionary biologists call an outgroup. Two pretty good outgroups for looking at bony fish, reptiles, and mammals are the primitive-looking hagfish and lampreys. The five bones to follow are the dentary, articular, squamosal, quadrata, and stapes bones. Because the bones have changed in the evolution of fish, reptiles, and mammals, the inner ear of these three organisms contains different structures. The inner ear also contains the balance organs, which are pretty well developed in fish, so we will also bring hagfish and lamprey balance organs into the discussion.

Hagfish and lampreys do not have jaws, but they do have inner ear vesicles. They feed primarily by grasping prey with their tongue or by attaching to their food item, so they get by without jaws fairly well. The anatomy of their heads with respect to the five bones is pretty simple. Since they don't have a jaw, they either don't have the five bones we are following, or it is too difficult to see them, and hence we simply cannot make a statement about their existence. The inner ears of these two fish have fairly simple structures, which suggests that the common ancestor of jawless and jawed vertebrates had an inner ear. But because the bones of the heads of these two primitive-looking organisms don't look like any of the bones in jawed vertebrates, all we can say is that the inner ear is an ancestral characteristic of both jawless and jawed vertebrates. It is in this inner ear area where both sounds and balance are managed.

The balance organs in the inner ear exist in the outgroups but in a modified form. The inner ear of a hagfish has only one semicircular canal, a tubelike structure responsible for balance. Near the semicircular canal in the inner ear of a hagfish resides a patch of neural cells that have hairs on them that act as balance organs. The lamprey has two semicircular canals and also patches of neural cells with hairs on them that act as the mechanism for mechanoreceptors in balance. More advanced vertebrates (bony fish, reptiles, and mammals) have three of these canals arranged like the three axes (X, Y, and Z) of a three-dimensional graph. That the semicircular canals are involved in balance is well known, and as discussed previously with respect to primates, a sense of balance is critical for survival. In some mammals, balance was not critical, and the evolution of the inner ear in these organisms presents biologists with yet another important proof of the occurrence of evolution and its importance in the natural world (box 5.1).

Fish attain balance by yet another structure called the lateral line. This balance system, found in most fish, is used to detect motion and vibrations in the external environment. The lateral line is made up of cells on the skin that are called hair cells. Motion or vibrations displace the hairs on these cells of the lateral line, and such displacement is converted into a neural signal to the brain through a mechanoreceptor mechanism. The lateral line is exactly what its name suggests—a line of hair cells extending from the gill covers to the tail. Although most fish use the lateral line hair cells to detect vibrations around them, the evolutionary process has modified the lateral line hair cells in many species to detect electrical impulses, too. So, in addition to detecting motion or vibrations around the body, these lateral line cells detect electrical impulses from other animals too. Of course, for these modified electroreception cells to be useful they need an electrical field to detect. So, there are two ways such fields are produced. First, some fish use these receptors to detect electric fields that they themselves produce. In a process called electrolocation, these fish use the hair cell organs to detect objects, somewhat the same way that echolocation works, only with electrical fields. Second, all organisms passively produce the other source of electric fields, because the nervous system of any organism is essentially producing electrical impulses all the time. Fish with electroreceptors use this passive source of electrical fields to locate prey and predators.

Electroreception is not unique in the animal world, because it ap-

## BOX 5.1 | VESTIGIAL BALANCE IN SLOTHS

By studying mammals that rely little on balance, researchers have examined the plasticity or expendability of the semicircular canals. One group of mammals that do not use balance to any great degree are the three-toed sloths. These mammals move extremely slowly, and as the authors of a paper examining their semicircular canals state, "Significant travelling distances, speed and agility are not part of their locomotor repertoire." Sloths do descend their tree top perches once in a while to defecate, but overall they do not need an acute sense of balance. In defense of Darwin, evolutionary biologist Jerry Coyne has pointed out that Darwin himself predicted that traits not under natural selection might show greater degrees of variation and even decay, leading to his ideas about vestigial organs. When the inner ears of several three-toed sloths were examined, Guillaume Billet and his colleagues discovered that in every conceivable way of measuring the structure of the semicircular canals, these sloths varied more than other mammals. This higher degree of variation is indicative of very relaxed natural selection on the structures. Coyne has pointed out that the beauty of the study is not in showing that vestigial organs occur, because evolutionary biologists can cite many great examples of vestigial structures. What is significant about the work is actually catching the semicircular canals on their way to becoming vestigial.

pears to have evolved several times. Perhaps the most notable case of this convergence can be found in the duck-billed platypus. This strange mammal, a member of a small group of mammals called monotremes, has an electroreception system that did not evolve from the lateral line but instead developed from cells in skin glands that have been converted through the evolutionary process into cells with free nerve endings. Monotremes in general have electroreception, but the platypus is the most impressive, with forty thousand of these modified skin cells arranged along the bill as stripes. Platypuses are pretty good hunters, and more than likely these electroreceptors assist in their acute hunting ability. Other vertebrates with electroreceptor capacity include some dolphins, and the anatomical origin of these electroreceptors is independent of the lateral line and also the monotreme adaptation.

The hagfish, as noted, does not have a jaw, so the next innovation in-

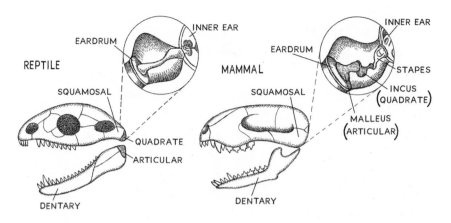

Figure 5.1. Comparison of bones of the jaw of a mammal and a reptile.

volves organisms with jaws. The anatomical structure we call the jaw is an ancestral characteristic of bony and cartilaginous fish, reptiles, birds, and mammals (fig. 5.1). The five bones we are following are also sometimes hard to find in jawed fish. In primitive sturgeons, the bones of the head are fused into a single sheet, so discerning the individual bones is extremely difficult. But when advanced bony fish are examined, the individual bones of the jaw become recognizable. In bony fish, we easily recognize and localize three of these bones—the dentate, articular, and squamosal bones–but there is no trace of the quadrate and stapes bones. Fish do not have an outer ear, nor do they have a middle ear, so it is safe to say that the three bones that are recognizable in fish jaws reside well outside of the ear. These two bones appear in reptiles, though, where the quadrate is found outside of the inner ear, and the stapes is found inside of the inner ear connected to a membrane called the eardrum. Note that the articulation of the jaw in bony fish is produced by the contact of the quadrate and articular bones, and this is a universal characteristic of jaws in fish, reptiles, birds, turtles, and amphibians, but not in mammals. What has happened in this last group results in the incredibly complex inner ear that mammals have. The quadrate and articular bones of the jaw hinge in mammals are converted into inner ear bones called the malleus (the same as the articular) and the incus (the same as the quadrate). This new arrangement of the five bones we have been following, especially the last three (articular, quadrate, and stapes or malleus, incus, and stapes), makes up the three-boned inner ear in humans and other mammals. The mammalian jaw is made up of the articulation of the squamosal and dentary bones. This is quite a transition, but ultimately it is a frugal use of the jaw bones. There has been and continues to be strong selection on maintaining these bone structures and articulations in mammals, and hence there is little variation in mammals' overall inner ear bone arrangement. But the range of sound wavelengths that can be detected by mammals can vary as a result of altering other structures of the inner ear, or even some of the structure of the middle and outer ear. One need only look at the size of the outer ear in mammals that require acute sound detection (larger ears are usually correlated with greater acuity of sound detection) to see how the outer ear can be altered as a response to selection for collecting and transmitting very specific sounds to the brain.

A modification of the inner ear of mammals that can produce variation in the range of hearing of mammals is the development of a coiled membranous extension at the bottom of the inner ear called the cochlea. Mammals are the only vertebrate with well-developed cochlear structures, and there is considerable variation in the makeup and length of this membranous structure. The cochlea varies in several aspects—total volume of the cochlea, number of spirals, total length of cochlea, diameter of the cochlear tube, tightness of the spiraling of the cochlear tube, and curviness of the cochlea—all of which have been correlated to the relative frequency of wavelengths that are detectable by the mammalian ear. For instance, mammals with large cochlear diameters are more sensitive to high-frequency sounds. So, one way mammals can adjust to the wavelengths that they can detect is through evolving different cochlear dimensions. The degree of variation of mammalian cochlear structures is impressive, reflecting the wide range of hearing capacity in mammals.

# 6  SUPERSMELLERS AND SUPERTASTERS

*The Limits of Smell and Taste in Humans*

"*I find people who devote their whole lives to taste a little strange.*"
—Jonathan Safran Foer, author

The variation in our senses is substantive and a good way to investigate how they work in humans. Some of the examples you're about to read could be straight out of *The Guinness Book of World Records* and indeed are more than likely listed in that entertaining record of human limits.

Joy Milne can smell Parkinson's disorder. When randomly handed six T-shirts worn by people with Parkinson's and six worn by unaffected individuals, she correctly identified who had Parkinson's for eleven of the twelve shirts. Not bad, but even better, because she actually achieved twelve out of twelve. The one she missed was a false positive: she identified one of the shirts from the control group as belonging to someone with Parkinson's, and this person was later diagnosed with the disease. The smell of Parkinson's was heavy on Milne's mind because her husband was showing more and more extreme symptoms of this devastating neurological disease. She developed her unique ability when her husband's condition worsened and she noticed that he exuded a musky odor. Sadly, her husband has since died from the symptoms of Parkinson's disorder. Although one might expect the odor to emanate from the armpits or some other sweaty region of the body, it actually comes from the sebaceous glands of the back, chin, forehead, and neck. These glands secrete a product called sebum that leaves a shiny veneer to the skin in the areas where it is secreted. Apparently, something in the sebum rubs off onto the shirts of the people that Milne smelled.

In another smell-related neurological disorder, evidence presented in 2015 suggests that diagnosing Alzheimer's disease might benefit from assessing the sense of smell. Alzheimer's is a brain disorder that usually has a late onset in humans. It is characterized clinically by loss of memory and confusion in early stages and by extreme neurological degeneration in later stages. People with the disorder develop large plaquelike structures in their brains; the plaques are thought to be one of the causes of the neurological problems associated with Alzheimer's. The disease is the only top-ten deadly illness that has no cure or efficient way to slow it down. It afflicts mostly people of western European ancestry. Forty-four million people worldwide have the disease, including five million Americans. In addition, seven hundred thousand people are estimated to die of the disease each year. Scientists have known since the 1980s that some, though not all, Alzheimer's patients develop a very poor sense of smell. In addition, mice fed tiny amounts of beta-amyloid, a protein found in the brains of people suffering from Alzheimer's, showed the formation of plaques in their brains, linking plaque formation to the beta-amyloid intake. When these mice were studied for their olfactory acumen, it was found that they spent more time sniffing around objects than normal mice and also could not remember odors. Given that mice have evolved to use smell to interpret their outer world, the loss of this sense correlated with the beta-amyloid intake is significant, and researchers started to look for similar phenomena in humans. A fascinating outcome of this study in mice is that the researchers next removed the beta-amyloid from the mice and the sense of smell returned.

Correlating the loss or alteration of the ability to smell with Alzheimer's is tricky. If one tests people with the disease for reduced capacity to smell and sees a correlation, then it wouldn't mean much for early diagnosis. It is crucial to catch the diminished capacity to smell before the onset of the disorder for loss of smell to be a good diagnostic tool. Researchers at several institutions in New York City developed a survey to do just what is needed to make the loss of smell a diagnostic before the onset of the symptoms. Nearly four hundred older people (averaging eighty years) without Alzheimer's symptoms were enrolled into a study and given the University of Pennsylvania Smell Identification Test (UPSIT). The test is basically a scratch-and-sniff affair with questions about the odors emanating from the scratch areas of a forty-page test. The answers were tabulated and compared with a panel of answers from four thousand control individuals with normal olfactory capacity. In addition, the 387 participants were examined with magnetic resonance imaging (MRI) to assess the thickness of the entorhinal cortex of the brain. This is the first part of the brain to be affected by the conversion to Alzheimer's, and so it is a logical place to examine for anatomical changes. The study participants were contacted again in a follow-up four years later; 20 percent showed signs of diminished mental capacity, and nearly 13 percent had developed Alzheimer's. The trick, then, is to go back to the UPSIT data and the MRI data to see if either correlate with the development of Alzheimer's disease.

Surprisingly, low smell-test scores, indicating diminished olfactory capacity, were strongly correlated with the development of Alzheimer's, whereas the thicker entorhinal cortex in MRIs was not. In another study of eighty-four elderly individuals, the UPSIT test was administered in an attempt to understand the loss of olfaction and its co-occurrence with Alzheimer's. This time, the research team, instead of taking MRIs of the brain, made positron emission tomography (PET) scans and analyzed the cerebrospinal fluid of these older adults (average age was seventy-one). The PET scan can detect plaques in the brain, and the cerebrospinal fluid can be analyzed biochemically to detect amyloid. Both are diagnostic of Alzheimer's. In a follow-up six months later, 67 percent of the participants showed signs of cognitive decline. But testing positive for amyloid (using either PET or cerebrospinal fluid) was the better diagnostic. On the other hand, participants who scored low on the UPSIT scale at a particular threshold were three times more prone to develop cognitive decline as those with scores above the threshold. Both studies support the idea that the UPSIT approach could be a good early indicator of the onset of this terrible disease. Low UPSIT scores as a diagnostic tool could lead to much earlier intervention to curb the progression of this debilitating disorder.

Smelling Parkinson's and not detecting smells well because of Alzheimer's are excellent examples of the range of human olfaction. In one case (detecting Parkinson's disease by smell), the individual increased her acumen for olfaction; she became a "supersmeller." In the other instance, individuals lose the ability to smell. All of this goes on in the nose, which then communicates with the brain. In fact, the information from the nose goes to and is processed in a small region of the brain called the olfactory lobe.

Consider Joy Milne. From photographs, she clearly has an ordinary-looking nose on the outside and more than likely has the same ordinariness on the inside of her nose (fig. 6.1). Externally, she has two nostrils through which the small molecules that constitute odors are inhaled. Internally, she probably has a perfectly normal nasal passage lined by

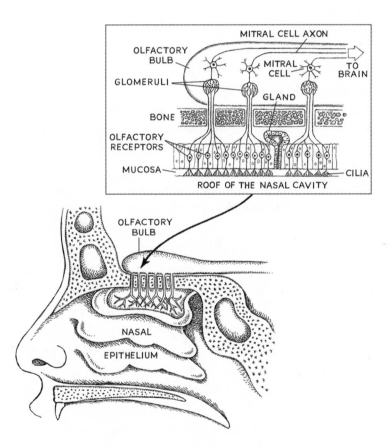

Figure 6.1. Structure of the nasal passage and nasal epithelium. *Inset:* the glomeruli and their connections to the olfactory bulb.

fine hairs to keep out large particles of dust and other junk in the air. Her nasal epithelium, where the initial olfactory action occurs, would also look very normal. Even viewing the nasal epithelium with an electron microscope would show two normal-looking kinds of cell, the first being stem cells that generate new nasal nerve cells throughout her lifetime. The other kind of cell is more complex, but these cells in Milne's nasal epithelium would look much like the cells in anyone's nasal epithelium.

There are millions of nerve cells in Joy Milne's nasal epithelium. On the end of each of these millions of nerve cells facing outward into the air that is passing through the nasal cavity are small, hairlike structures called cilia. The cilia have a lot of mucus around them and are kind of swimming in it. On the end of the cell pointing inward are projections called axons that run directly to the specific part of the brain called the olfactory bulb. In this bulb, which extends off the brain underneath the frontal cortex and pretty close to the nose, there are twenty-five or so cells that bundle the axons. Each one of these cells, called mitral cells, can have up to twenty-five thousand axons running through them. Inside the mitral cells there are small microscopic bodies called glomeruli, where the nerve axons congregate to form hubs and ultimately connect the cell to the brain. The glomeruli can bundle multiple axons, and Milne's glomeruli and olfactory would look pretty normal, too.

But the olfactory bulb isn't the end of the line for the information bundled into the glomeruli. Neurons travel out of the glomeruli and connect the bulb to the olfactory cortex, which is located in the cerebral cortex at the boundary of this part of the brain with the temporal lobe. It is this pathway from the bulb to the cortex that researchers think is responsible for storing memories about odors. In essence, it is probably where Marcel Proust's famous madeleine cake memory (box 6.1) resided in his brain. But the information from the cilia travel a little farther into the brain to the orbitofrontal cortex, where the information from olfaction is integrated with other higher brain functions. Although what Joy Milne's brain looks like is not public information, I would bet that she more than likely has a pretty normal olfactory bulb, olfactory cortex, and orbitofrontal cortex.

So what is different about Joy Milne's olfactory apparatus that gives her the supersense? Perhaps by looking at the odor itself, that musky smell she detected, we can shed light on her supersense and, in the pro-

The following quotation from Marcel Proust's *Remembrance of Things Past* is the origin of the oft-quoted and referenced madeleine, a small cake baked in the shape of a scallop shell. The madeleine has become a sort of metaphor for stored memories, how vivid and wonderful they can be, and how they can be triggered by simple sensory stimulation such as smell: "She sent out for one of those short, plump little cakes called petites madeleines, which look as though they had been moulded in the fluted scallop of a pilgrim's shell. And soon, mechanically, weary after a dull day with the prospect of a depressing morrow, I raised to my lips a spoonful of the tea in which I had soaked a morsel of the cake. No sooner had the warm liquid, and the crumbs with it, touched my palate than a shudder ran through my whole body, and I stopped, intent upon the extraordinary changes that were taking place. . . . At once the vicissitudes of life had become indifferent to me, its disasters innocuous, its brevity illusory."

cess, on how the sense of smell works. What causes the Parkinson's odor is a mystery. The culprit could be a small protein called alpha-synuclein that is important in the expression of Parkinson's disease. This protein is a whopping 140 amino acids long (which is a couple orders of magnitude larger than most molecules responsible for odors), and its three-dimensional structure resembles an unbent paper clip. Alpha-synuclein is also responsible for loss of smell in people with Parkinson's because it forms clumps in the olfactory bulb. Because of this clumping in the olfactory bulb and other regions of the brain, it was immediately suspected of being involved in the musky odor that Milne detected. But the odor itself emanates from the sebum and could be some other smaller molecule.

In fact, musky smells have been used by humans for a long time and are so well understood that most musky-smelling solids or liquids are now made synthetically. More than likely, the first musk was obtained from musk deer. It comes from a sac that looks much like a scrotum on the underside of male deer of this species. (The word "musk" is actually derived from the Sanskrit word for testicle.) Anyone who has been around musk deer or beavers can attest that natural musk is not a pleasant-smelling compound. It needs to be diluted and treated with alcohol for it to attain its pleasant odor—otherwise it is pretty foul-smelling. Many molecules are responsible for musk odor. All are part of the liquid that the musk deer stores and eventually sprays from the musk sac. Most of these molecules contain seventeen or eighteen carbon atoms (as opposed to the huge alpha-synuclein's more than twelve hundred carbon atoms), sometimes arranged in an irregular ring. They have much smaller and very different shapes than alpha-synuclein.

And the shape of the musk molecule is everything when it comes to how the nerve cells in the nasal epithelium work. Embedded in the membranous part of the cilia in the nasal epithelial nerve cells are relatively large molecules appropriately called odorant receptor proteins. These proteins are securely embedded in the membrane because the protein loops in and out of the membrane seven times. It starts on the outside of the cell membrane and goes in, out, in, out, in, out, and finally back into the inside of the cell. The protein itself has two "business ends": one on the inside of the cell and one on the outside. The part of the protein that sticks out of the membrane is shaped by the amino acids that are at that end of the protein. And the shapes are unique from one odorant receptor to the next. Remember that mammals have varying numbers of these odorant receptor genes in their genomes. Our species, for instance, has about a thousand total genes, but only about four hundred of them actually work, putting the other six hundred genes in the pseudogene category. So, we have four hundred differently shaped odorant receptor proteins that weave in and out of the membranes of the many cilia in our nasal epithelium, all of them having different shapes sticking out of the nerve cell. Each odorant receptor cell has a unique complement of proteins and hence a unique complement of the different shapes sticking out of the cell. The business end of the protein on the outside of the cell is what recognizes the odorant molecule. But the big question is how.

Most biologists would immediately point out that the musk odorant simply collides with a receptor into which it can fit and with which it can interact with a type of lock and key or, as some call it, a hand-in-glove mechanism. Once the odorant molecule fits in with the odorant receptor protein, this changes the three-dimensional structure of the receptor on the end of the protein inside the cell, which in turn triggers reactions in the interior of the cell.

The biophysicist Luca Turin, for his part, made the interesting and unorthodox suggestion that the odorant molecule (in our case, the musk odorant) does indeed fit into the receptor protein, but instead of changing the conformation of the protein on the interior of the cell, it does something very different. Turin's idea is based on the vibration of molecules, because the electrons in the atoms of the protein are moving around as part of chemical bonds. Imagine a chemical bond where two atoms share an electron. As the electron moves from one atom to the other as a result of the bond, it will twitch or vibrate. The vibrational theory suggests that when the odorant molecule fits into the receptor, it changes the vibrational property of the protein. This vibrational change allows electron transfer to the receptor. The electron then travels from one end of the receptor to the other, which is the cause in the change of the vibration of the receptor. The electron transfer or change in vibrational frequency eventually triggers a cascade of events in the interior of the nerve cell.

Some researchers and journalists have called this the "swipe card theory," as opposed to the more orthodox lock-and-key mechanism favored by many olfactory researchers. It is an idea that is actually nearly 150 years old, because an unnamed scientist proposed this hypothesis in 1869 in *Scientific American,* the premier American science journal at the time. Turin's role in the resurrection and development of the theory is substantial, however, and has led to the development of some clever experimental approaches (box 6.2). When these tests are done, some of the predictions of the vibrational hypothesis bear out, but some don't. Although the results of the tests are suggestive, they are very difficult to interpret, as Lesley Voshall and Andreas Keller have pointed out, because the experiments that lent credence to the vibrational theory have been challenged. In 2015, Eric Block and colleagues performed experiments with specific receptors, one from a human and one from a mouse, using the deuterium approach (box 6.2). They took isotopomers of the musk odorants and tested them for olfactory differences. None were detected, and so this would reject the vibrational hypothesis.

So, the vibrational theory remains an interesting idea, but the hand-in-glove theory based on shape of the odorant and its receptor pocket seems to be more substantiated. Whether the vibrational theory or the shape theory or perhaps a mixture of both wins out, the same process has to happen once the odorant interacts with the receptor: a cascade of

## BOX 6.2 | TESTING THE VIBRATIONAL HYPOTHESIS

Several very clever experiments have been devised to test the validity of the vibrational hypothesis. These approaches involve using deuterium, an isotope of hydrogen. Deuterium has a neutron in its nucleus compared with hydrogen, which has none. A compound made using deuterium will have different properties than the same compound made with hydrogen. One of the different properties of a compound incurred by using deuterium is the vibration of the molecule. By substituting deuterium for hydrogen in an odorant molecule, one can significantly alter the vibrational property of the odorant, making the molecule with deuterium what is called an isotopomer of the one with hydrogen. The size and shape of the odorant are hardly altered by the insertion of deuterium for hydrogen, but the vibrational properties are. If the vibrational theory is correct, then the odorant with the deuterium should smell different from the odorant with hydrogen. Validity of the shape hypothesis would lead to both smelling the same because their shapes are the same. Another test would be to find two odorants with the same vibrational properties but different shapes. In this case, if the vibrational theory is correct, then the two odorants should smell the same, as opposed to the shape theory, where the two odorants should smell different.

protein interactions inside the cell occur that create an action potential, the currency of the nervous system. This impulse in turn transfers the initial information about the odorant to the brain. For several of the senses this mechanism is the same. The process itself is called signal transduction, and it involves a protein complex that I discuss in the context of the other chemosensory sense—taste.

Taste has many fewer genes that code for receptors in our genomes than for olfaction. Although there are about four hundred functional olfactory receptor genes in humans, an order of magnitude fewer are taste receptors. This small number of taste receptors doesn't mean that the range of variation in what different humans can taste is narrow, however. It is important to consider how the small molecules that produce taste interact with taste receptors before seeing how broad the taste receptors are. Some of these interactions are quite different from how odorants interact with their receptors.

Cells that detect taste (for that matter, most cells) are fairly complex entities. Their membranes are littered with small molecules, including the receptor proteins that have already been described. Many of these proteins are securely anchored in the membrane by means of varying numbers of loops of the protein in and out of the cell. Another kind of protein embedded in the membrane is called an ion channel. This protein does what its name suggests, by transporting ions (atoms with electrical charges) from the outside of a cell to the inside and vice versa. Other proteins are embedded in the membrane but are of less importance in the recognition of taste or smell. Inside the cell, there is the obligatory nucleus and other organelles that keep the cell working, such as the mitochondria, where energy for the cell is processed, and the endoplasmic reticulum, where proteins are synthesized. But these cells also include small bodies or sacs called vesicles that congregate at the part of the cell where the nerve impulse will be transferred to a neuron for eventual connection to the brain. The vesicles are chock-full of small molecules called neurotransmitters that are integral parts of transmitting electrical messages from one cell to the next. The electrical messages are called action potentials.

As discussed in Chapter 4, taste receptors recognize five major categories of "tastants"—salty, sweet, bitter, sour, and umami. Recognition of each of these five taste categories is implemented by a different kind of small molecule or even part of a molecule called an ion. For instance, table salt (sodium chloride, or NaCl) has two components: a sodium atom that is missing an electron and a chloride atom that is missing a positron, making the sodium atom positive (Na+) and chloride atom negative (Cl−). The two ions are loosely connected by sharing what is called an ionic bond that is pretty weak as chemical bonds go. Saltiness is recognized when the positive ion of a salt (Na+ in the case of table salt) is transported across the taste cell's membrane via the ion channels discussed above. When a salt like sodium chloride congregates around taste buds, the outside of the taste bud cells is inundated with Na+ ions. These Na+ ions are rapidly transported across the cell membrane in the ion channel, that small, porelike machine that pumps the ion across the membrane. When a sufficient amount of Na+ ions collect in the nerve cell, it does a trick called depolarization, in which it sucks calcium into the inside of the cell via ion channels. The calcium atoms are doubly positively charged (Ca++), and they induce the vesicles to release their contents. Loads of small molecules called neurotransmitters are released into the area between the taste cell and the adjoining neural cell, called a synapse. The cell needs to reset itself, or void itself of all of the positive ions inside, so once the vesicles have done their job, the ion channels back-transport all of the potassium on the inside of the cell. The potassium ions (K+) are positive, and this resets the charge on the inside of the taste cell, creating an electrical charge or action potential that again is the currency of the nervous system. Acid recognition is accomplished in a similar manner as salt detection.

Because all acids have one thing in common—a weak bond involving hydrogen that produces hydrogen ions (H+)—it is a hydrogen ion that triggers the ionic changes on the inside of the taste cell. You might be asking, if it's the same mechanism, then why isn't acid just a salty taste? It turns out that the H+ ions also block the movement of K+ ions across ion channels and also enhance the entrance of other positive ions into the cell. Hence, the salty and acidic tastes are created by different kinds of ionic changes that differentiate the two tastes. In addition, with acidic compounds, the cell vesicles recognize this different kind of accumulation of positive ions, and only those vesicles that should respond to H+ are released. As with salts, the cell needs to reset itself, so after vesicle release, the potassium channels are cleared and the K+ ions on the inside of the cell are transported out for the reset.

Unlike salty and acid, the tastants for sweet, bitter, and umami do not enter the cell. Instead, they interact with receptor proteins embedded in the cell membrane, much like the odorant receptors described previously. These receptor proteins interact with what are called G protein complexes (fig. 6.2) that are positioned on the inside of the cell (box 6.3). The G protein–coupled cascade will also activate the vesicle release.

The vibrational theory could also work for taste. As with smell, this general vibrational idea is not new, because that same 1869 *Scientific American* article mentioned previously proposed the same hypothesis for taste. The recent advances in the potential of the vibrational theory are based on the knowledge of molecular biology that simply didn't exist in the nineteenth century. Nonetheless, the way that G protein complex receptors work leaves open the possibility that some receptors work better than others, and there is enough variation in human populations for

The G protein complexes have three subcomponents that are bound to one another: alpha, beta, and gamma. This three-protein complex is connected to the actual receptor molecule described as having two business ends. One end is on the outside of the cell and can bind to the odorant molecule or, in the case of taste, the tastant molecule, in a hand-in-glove manner. The other end is on the inside of the cell and interacts with the G protein complex. Once the odorant or tastant molecule binds to the outer business end of the receptor protein of the smell or taste cell, the internal end of the receptor induces a reaction in which the G protein subunits are cleaved into an alpha-only protein and a beta + gamma–complexed protein. These two proteins activate other proteins inside the cell to induce vesicle release into the synapse. Each of three tastes—sweet, bitter, and umami—induce the vesicle opening into the synapse in different ways, resulting in the differentiation of the three tastes by the brain.

these receptors that there is quite a range of tasting for humans. In addition, the numbers of receptors that exist on the tongue can have a huge impact on human variation for taste.

In general, most humans can be placed into three major categories of tasters—nontasters, tasters, and supertasters, roughly in the ratio of 25 percent:50 percent:25 percent. There is also a small percentage (less

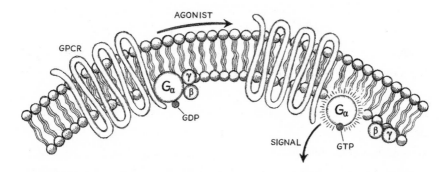

Figure 6.2. G protein–coupled receptors (GPCRs). The GPCR spans the cell membrane seven times and complexes with alpha, beta, and gamma proteins on the inside of the cell. When an agonist (a neurotransmitter) interacts with the protein on the outside of the cell, this cleaves the gamma and beta proteins from the alpha protein and converts GDP (guanosine diphosphate) to GTP (guanosine triphosphate), which then triggers a signal in the cell.

than 1 percent) of humanity categorized in a super-supertaster category. Supertasters are mostly women, and people of European ancestry are usually not supertasters. So what exactly is a supertaster? You might think that a supertaster would have a lot of fun eating and drinking, but it's more like the opposite. Because supertasters experience tastes more intensely than nontasters and tasters, the effects of different tastes detected by tongues of supertasters are amplified relative to the nontasters and tasters. Super-supertasters have it even worse than supertasters. Taste is a good case of "more is *not* better."

The best way to describe the differences between the categories of tasting is to take one of my favorite beverages to taste—beer—and explain how each of the categories of tasting will respond to this beverage. The Master Brewers Association of the Americas recommend what is called the American Society of Brewing Chemists flavor wheel to help its members assess the taste of their brews. The flavor wheel was created by a coauthor of *Sensory Evaluation Techniques,* first published in the 1970s and now in its fifth edition. Morten Mielgaard, a professor of the senses and how to measure them, created the taste wheel to lend a more quantitative aspect to beer tasting.

The taste wheel is quite complex and has gone through many iterations since Mielgaard created it, but it does focus on the complexities of the perception of beer. Examples of the more than one hundred possible categories of taste include grapefruit, caramel, farmyard, funky, burnt tire, and baby sick/diapers (which I hope never to taste). It is safe to say that these tastes are the result of many factors, but they all emanate from the very simple contents of beer. In fact, to protect the simple contents of beer, in 1516 Germans created the Bavarian Beer Purity Law, or Reinheitsgebot. The purity law forbids any beverage labeled "beer" to be made with anything but hops, water, and barley. Although yeast is needed in brewing, it is a microbe, and was obviously not recognized as an ingredient five hundred years ago. So, the modern concept of taste in most classical beers comes from only four ingredients. The most interesting aspect of the taste of beer, at least to me, comes from the hops and the sugars in the brew, and of course the alcohol that is the product of fermentation implemented by yeast on the sugars from grain.

Although beer is probably several millennia old, hops have been a part of brewing beer for a little more than a millennium. Its widespread use began in the last eight hundred years in Germany and was cemented in brewing technology with the invention of India pale ale (IPA) in the early to mid-nineteenth century. With the modern advent of microbreweries and the development of custom-made hoppy beers such as the many IPAs that are on the market, this beverage becomes one that has a wide range of bitterness. It might be surprising to note that hops were first used as a preservative in beers. The bitter taste from hops is an afterthought. The manipulation of hops today as an integral ingredient in producing craft beers makes for some pretty wildly hoppy beers. (All of which I enjoy immensely, making me more than likely a normal taster.) Supertasters find beer incredibly bitter, so much so that they will avoid drinking hoppy beers like IPAs and will not be too terribly enamored of even mildly hopped beers, like most lagers. I also am immune to the burn of alcohol in the beer, something that a supertaster will report when his or her lips touch a high-alcohol beer. Suffice it to say, hard liquor is a no-no for supertasters. Nontasters will pretty much drink and eat anything, so their tolerance for hoppiness is extreme. But they more than likely will not be able to tell the difference between a Columbia hopped beer and a Cascade hopped beer. Supertasters more than likely should be able to discern quite well between these two hops by taste, but unless they have been conditioned to drink beer, they more than likely will first and foremost consider both as just really bitter. So it is normal tasters who have all the fun with tasting hoppy beer. All of this does not mean that supertasters and nontasters won't enjoy alcoholic beverages. A nontaster will have no trouble gulping down a jalapeño-infused tequila, and a supertaster can be conditioned to drink beer or wine and enjoy those beverages. It has been suggested that upscale chefs are supertasters who have conditioned themselves to overcome the overwhelming effects on their taste buds and to use their supertasting as a tool to create novel dishes. Recently, sour beers or farmhouse ales have become very popular. In this case, brewers of these interesting beers take advantage of the sour taste receptors and combine that with a little hoppiness. Anyone who has tasted a really sour farmhouse ale, though, will recognize that their sour taste receptors are going crazy relative to their bitter receptors.

Tasting beer is really a simple reaction of the small chemicals in the brew with receptor molecules on the tongue. But unlike smell, and although taste is combinatorial, the one thing that dictates whether someone can taste, supertaste, or not taste is ultimately caused by the number of taste cells on the tongue. The taste receptor cells are found in bundles of anywhere from thirty to one hundred cells within which the taste receptor proteins reside. The bundles of cells are called taste buds, and most of these reside on physical structures of the tongue called papillae.

There are three forms of papillae on the tongue related to where they reside. The fungiform papillae reside on the anterior region of the tongue, look like little mushrooms budding up from the surface, and can have up to two taste buds per papilla. The circumvillae are located in the posterior region of the tongue, and the foliate papillae are located on the sides of the tongue. Taste papillae are also found on the upper side of the mouth (the palate) and in the throat. Taste cells have also been found in the lungs, but their function in this tissue is unknown.

The density of papillae on the tongue is directly correlated to being a supertaster (more than thirty per 100 mm$^2$), a taster (fifteen to thirty per 100 mm$^2$), and a nontaster (less than fifteen per 100 mm$^2$). In this case, instead of the genes for the taste receptors being the ultimate cause of supertasting, an underlying developmental process is involved. How the tongue develops its papillae has recently been deciphered, and interesting hypotheses about the evolution of the arrangement and number of papillae formed during development are being tested. One immediate result of this work is the discovery of the strange phenomenon that teeth and papillae are patterned with similar genes.

What is an effective technique for examining how many papillae someone has in a given area of the tongue? All of them involve darkening it, and the most enjoyable is to swirl red wine in the mouth and over the tongue. If done correctly, you will be able to see little lumps of tissue on the tongue that are the papillae. Next, take a piece of three-hole notebook paper. The punched holes are about 6 or so millimeters in diameter, and a piece of paper torn off with one of these holes can be placed over the darkened tongue. Now simply count the number of papillae you see in the punched hole. If you have fewer than fifteen papillae, you are more than likely a nontaster, whereas from fifteen to thirty papillae would suggest that you are a taster. Anything over thirty would indicate that you are a supertaster or a super-supertaster (fig. 6.3).

Believe it or not, the reception of tastes is somewhat similar to how pain is perceived. In fact, one of the best ways to describe how pain re-

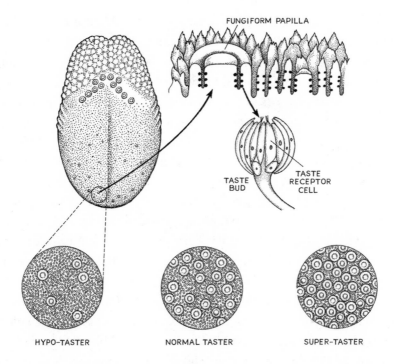

FUNGIFORM PAPILLA

TASTE BUD

TASTE RECEPTOR CELL

HYPO-TASTER        NORMAL TASTER        SUPER-TASTER

Figure 6.3. Tongue papillae and the types of tasters in human populations.

ception works is to look at foods that have tastes that are painful, such as spicy, hot foods.

He has bathed in beer, chocolate (apparently milk chocolate), and fifteen hundred pureed Oreos, but his latest bath was a doozy. Cemre Candar, an Internet sensation (whatever that is, and whatever it gets you) has more than fifteen million hits for some of his stunts, and his feat in 2016 drew more than two million viewers over the first week or so it was posted. You see, he completely immersed himself in hot sauce. If that weren't enough, he topped the bath off with a bucket full of red-hot chili peppers. I don't recommend doing this or even watching the video of this stunt. It looks excruciatingly painful.

Cemre Candar is indeed a strange human, and his bath in hot sauce was, as he can tell you, very painful. But why? It's just a liquid, isn't it? But the hot sauce and hot peppers, while made up of a lot of water and salts, also contain a small molecule called capsaicin. This small chemical found in many peppers reacts with and transduces specific cells in our bodies. The amount of capsaicin in the pepper or hot sauce dictates how hot the sauce actually is. There is a measure for the heat that a pepper can emit, and it is called a Scoville heat unit (SHU). It's determined by undertaking a little organic chemistry on the pepper, and then feeding decreasing concentrations of the organic extract to a panel of experts until a majority of the panel can no longer detect the heat. The concentration at which the heat can no longer be detected is called the heat level. To demonstrate the scale, consider a jalapeño pepper. If you think these peppers are hot, watch out for the Carolina Reaper (or, rather, stay away from it). A typical jalapeño has a Scoville rating of more than 2,500 SHUs. The hottest Carolina Reaper on record had a rating of 2.5 million SHUs, or more than ten thousand times hotter than the jalapeño. It's hard to tell from the video, but more than likely, Cemre Candar used a Tabasco-style sauce to fill his bathtub (he used 1,250 bottles of it), topped off by whole red habanero chilis. Tabasco (depending on what variant you use) has a rating of about 5,000 SHUs, and red chilis also rate about 5,000 SHUs. If he had used anything like Blair's 16 Million Reserve hot sauce, the bath would have been incredibly expensive (this is a top-of-the-line product), and way hotter, as this sauce comes out of the bottle at 16 million SHUs. Even with the relatively mild Tabasco and red chilis, his bath would have been quite concentrated with capsaicin. So, what happened to Cemre Candar's body as he slowly lowered himself into the red concoction in his bathtub?

As he lowered his body into the tub, the small capsaicin molecule bathed his skin. The skin has all kinds of cells but some also have a transient receptor protein (TRP) embedded in the membrane (the one greatly affected by the hot sauce bath is called TRPV1). This protein is a little like the chemoreceptor proteins discussed previously, except instead of weaving in and out of the membrane seven times, like the other chemoreceptors, the TRP makes only six turns. The big difference, though, is that instead of having ends flapping on the outside and inside of the cell like an olfactory receptor, the TRP makes a channel in the membrane with its six transmembrane domains that is critical for its functioning.

High-temperature, low pH (acid), and small molecules like capsaicin will activate the channel. A compound in hot mustard and wasabi called allyl isothiocyanate will also activate the channel. Each of these insults the skin and will open the channel so that the opening can then regulate the external Ca++ and Na+ (and their intercellular counterparts) concentrations. Without going too deeply into the neurochemistry, we can say that this chemical regulation affects the voltage regulation and is at the heart of the action potential that then sends the information to the brain via the nervous system. In Cemre Candar's case, the capsaicin hit the TRPV1 channels and opened them up wide, forcing the cell to pump Ca++ back and forth to regulate the concentration of this molecule. This triggered a regulation of voltage in the cell that then ran through Cemre Candar's nervous system as action potential to his brain, where the response in his brain was one of pain. The further he immersed himself in the red goo, the more of his cells were opening their TRPV1 channels and the more action potential went rushing to his brain to express pain. Oh, and he also realized that it was hot, because the TRPV1 channels would also relay that message to the brain. Although we can't see him sweat in the red goo, it is well known that a physiological response to overexcitement of TRPV1 channels by capsaicin is sweat, and he was probably doing this profusely while in the tub. The sensing of pain in this case was caused by a silly penchant for the extreme, but there are unfortunate cases where the loss of sensing pain can be injurious.

In 2006, six children were brought to the attention of medical researchers in Pakistan with strange healed and unhealed injuries. Not all of the children were related, but three were from one family and two were from another single family. Their injuries were bizarre, because the children had never complained of being in pain when the injuries occurred. Several of the children were missing the tips of their tongues, having bitten them off in early childhood without even a whimper. The older children of the six did learn to feign being in pain when injuries to them looked particularly bad. Almost all had at least one limb that had been broken and healed without the child mentioning the break to a parent. These remarkable children simply could not feel pain. On the other hand, they could perceive touch and temperature and were ticklish. Because there was a familial pattern to their lack of pain, researchers hypothesized that the problem in the six children was caused by the same phenomenon. In addition, because it was only pain that they could not perceive, scientists hypothesized that a specific kind of receptor in the children was not functioning. James Cox and his colleagues were able to map the loss of pain receptor to a specific location on chromosome 2 because they had family histories and samples of DNA from the families. The researchers then cloned a large chunk of the region where the lesion was localized and examined the chunk for genes that might be related to neurological function in general and pain reception specifically. They focused in on a gene called SCN9A, which codes for a sodium channel in the nervous system. Indeed, when Cox and colleagues examined the SCN9A gene of the six children, instead of finding a single mutation responsible for the lack of pain, they found three different genetic changes in the three different families that produced truncated genes. These truncated genes were effectively nonfunctional and resulted in the lack of functional pain receptor sodium channels in these six children.

In the decade since this study, major work on the genetics of pain receptors in humans has been accomplished. The disruption of normal pain reception and its transmission to the brain has been found to be incredibly complex. Many cell functions are disrupted in anomalous pain reception, including pathways involved in regulation of serotonin, estrogen, GABA, glutamine, and catecholamine. In addition, growth factors and other important proteins in development are involved. It appears that there are many flavors of pain that are mediated by various chemoreceptive and ion channel mechanisms.

# 7 WHERE AM I?

*The Limits of Hearing and Balance in Humans*

*"One person's roar is another's whine, just as one person's music is another's unendurable noise."* —Henry Rollins, musician

Anyone who has ever woken up after a rough night of drinking alcoholic beverages has likely experienced the room spinning viciously. In a flurry of chatroom posts over a fourteen-hour period in May 2006, several tech nerds discussed the phenomenon and proposed "cures" for it. Here is a sample of the banter: "The only advice I can give you is to wedge yourself into a corner of your room and hold onto the walls for dear life. Then phone your folks and tell them that the world, does indeed, revolve around you." Although overindulging in alcohol is no joke, it does serve the purpose of explaining how balance works in humans. In Chapter 5, I described the structure of the inner ear. Some of that structure exists for hearing, but the semicircular canal structures in the inner ear are there for balance.

These structures form a kind of X-Y-Z three-dimensional coordinate system known as the vestibular system. Each of the three semicircular canals has an area called an ampulla, where the semicircle of each canal meets. The canals themselves are filled with a fluid called endolymph. Inside each of the three ampullae is a gelatinous cellular structure called the cupula, which has small cilia, or hairs, emanating from its surface. These cilia are enervated and connected to the brain. If you rotate your head, the fluid moves with the rotation as a result of the inertia induced by head movement. Each cupula will lag behind like a floater on a fishing rod and will move in the opposite direction, which induces bending of the cilia in the ampulla. The bending hairs trigger an action potential in the canals, which is transmitted to the brain, where the information is interpreted to help us keep our balance. But balance isn't just about where our heads are in space.

There is motion all around us, and we are continually moving. In addition, how we perceive where we are in space involves a lot of random motion, called Brownian motion. If all of this random Brownian motion were detected by the vestibular system, it would cause a good deal of chaos in how we balance ourselves, because it would be transmitted to the brain as false information about our position in space and more or less overload our brains. Mees Muller and colleagues developed a model to examine how our semicircular canals overcome Brownian motion effects, and it involves several structural factors of the hair cells on the cupula. Specifically, these hair cells are ten times longer than the hairs in the auditory system, ten times less compliant to bending (the cochlear hairs of the auditory system bend ten times more easily), and one hundred times harder to displace than the cochlear hairs. They postulate that the strange X-Y-Z format of the vestibular structure is a good mechanical way to overcome the effects of Brownian motion on the sense of balance (fig. 7.1).

All of this information from the semicircular canals is transmitted to the brain and integrated with information from two other sources: the eyes and the muscles and joints. The integration of these three sensory inputs is transmitted to the brain stem, where it is integrated and interpreted by further contact with the cerebellum and cerebral cortex. These two areas of the brain are important, because they help the rest of the body respond to the initial movement that originated the need to balance oneself. The cerebellum is important because it coordinates complex movement of the body, and the cerebral cortex is important because it provides information from memory and learned experience to "right the ship."

Figure skaters are among the best balancers in human populations.

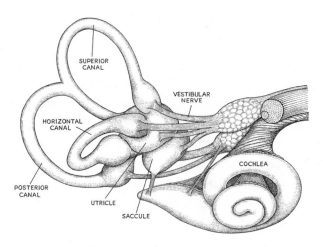

Figure 7.1. Structure of the inner ear. The balance organs are on the left, and the hearing apparatus is on the right.

They put themselves through incredible spinning routines that wreak havoc on their vestibular systems. For instance, as they start their spin, the cupula will, because of the inertia of the endolymph, move rapidly in the opposite direction of the spin. The cupula's cilia, or hairs, will be bent, indicating to the brain that the head is moving in the direction of the spin. But as the spins increase and the hairs get more and more bent because of the spin, the body of the skater, with the visual system and muscular movements, begins to respond. Even with this extreme perturbation of the cupula, whether it be by the classic scratch spin, the Biellmann spin, an "I" spin, a camel spin, or a pearl spin, skaters are amazing at pulling spins off without falling flat on their faces. Figure skater Natalia Kanounnikova set the world record for spins per minute at 308 rotations per minute in an "I" spin and skated straight away evenly as if nothing had happened.

How do skaters like Kanounnikova maintain their balance and deal with dizziness? Some of it is innate ability, because figure skaters are incredible athletes. But a lot involves using tricks to overcome the vestibular system's quirkiness. One of the tricks in spinning is the placement of the feet and hence the placement of the central axis along which the skater rotates. If a skater keeps this axis very tight and constantly upright, and maintains the center of gravity along the spin axis, he or she will have enough support to remain upright. Of course, the technique takes years of practice to perfect. Another trick is simply to execute a superfast spin at the end of the skating program. Here the skater can recover balance quickly if the spin ends in a very stable (but dramatic) pose. Another trick is to come out of a spin and go into a wide arc during which the skater can recover whatever balance that has been lost. Although the arc is often considered a poetic and artistic move, it really is a matter of survival that the skater uses to stay balanced.

With respect to someone who feels dizzy after a night of heavy drinking, if too much ethanol is ingested into a person's body and isn't detoxified by the liver, it will remain in the bloodstream, where it will travel to all sorts of places in the body. One place where it ends up in sufficient amounts to affect physiology is the inner ear and especially the semicircular canals there. The cupulae in the semicircular canal aren't used to being bathed in ethanol, and so when they are, they respond by distorting their shape and forcing the cilia to remain in constant contact with the sides of the ampullae. The nerve cells connected to the cupulae then send impulses to the brain indicating something is wrong with balance, and the brain responds by trying to compensate for the faulty information. Specifically, the brain starts the visual system spinning, and hence when someone drinks a little too much, the room might appear to be spinning. But it is also possible to get lucky and fall asleep without further discomfort. During sleep the ethanol will evaporate, and the cupulae will return to their original shape. But on waking the person might find the room still spinning. This spinning is the result of the brain remembering the bad behavior from the night before, thinking that it is spinning and setting the visual system spinning again to compensate for the memory.

Balance is one of those attributes where we more often detect defects in the sense, instead of superhumans who are very good at it. Our understanding of human variation in balance is based on anecdotes and the study of defects that certain humans have with respect to balance. Perhaps the most famous anecdotal instance of superbalance is the mythical case of Native American skywalkers (box 7.1).

And many recognized medical anomalies are rooted in the vestibular system, according to the National Institute of Deafness and Other Communication Disorders, part of the National Institutes of Health. Vertigo, for example, is a very disarming condition. It can be caused by several anatomical and physiological dysfunctions and goes by several

## BOX 7.1 | SKYWALKERS?

There are many anecdotes about the super balance characteristics of different ethnic groups. Mohawks from the Iroquios tribe have long been thought to be expert balancers, and this assumption has helped them find work in high places such as skyscraper construction sites and earned them the name skywalkers. In fact, in the New York City area, nearly 10 percent of the ironworkers on skyscrapers are individuals of Iroquois ancestry. But the myth is actually backward. The Mohawks of Kahnawake who were considered the original skywalkers lived near a railroad bridge construction site in the late 1880s. The ironworking company doing the construction hired several Mohawk men to work on the bridge, and they were quite successful because apparently they showed no fear of heights. It turns out, that is exactly what was going on. Although they showed no fear, they later admitted that they were very scared of the heights but simply did not show it as part of their cultural response to danger. Because to a Mohawk male, the career of a skywalker was and still is very warriorlike, the tradition of Mohawk and other Iroquois working in iron on skyscrapers was established and exists to this day. The skywalkers are more than likely no better at balancing than other groups of people are.

names. Benign paroxysmal positional vertigo (BPPV) and Meniere's disease are two of the more common disorders that result in vertigo. Many of the problems with the vestibular system are shared with the auditory system because of the nearness of the two apparatuses. In BPPV, structures in the auditory system called otoliths come loose from the saccule of the inner ear and enter the semicircular canals. These small, stonelike particles are important for hearing, but in the semicircular canals, they push on the cupulae. If the cupulae don't compensate, then cilia on them will be bent. This state will cause the neural cells of the inner ear to send false information to the brain about the position of the head. And of course, the brain will attempt to compensate for the false information by tweaking the visual system. The cause of Meniere's disease is not fully understood, but it is known to involve changes in the volume of the endolymph fluid in the canals.

Labyrinthitis and vestibular neuronitis can be caused by viral infections. Labyrinthitis causes parts of the inner ear to swell and causes loss of balance because the inflammation changes the relation of the cupulae to the ampullae. Vestibular neuronitis is an inflammation of the nerve leading from the vestibular apparatus called the vestibular nerve. Apparently, the inflammation of the nerve interferes with the positional information coming from the cupulae on their way to the brain. Like BPPV, the disorder known as perilymph fistula can be caused by head injury. The head injury in perilymph fistula causes the fluid of the middle ear to leak into the semicircular canals, which disrupts the positioning of the cupulae and results in lack of balance and dizziness. Those readers who have been on long cruises will be familiar with mal de débarquement syndrome. This syndrome usually manifests itself after being on a boat for long periods. Apparently, the vestibular system compensates for the up-and-down motion of a ship sailing on the ocean and continues to compensate after reaching land.

Balance has also been characterized across ethnic groups and age groups through the National Health and Nutrition Examination Survey (NHANES). These surveys, which began in the 1960s, have been conducted on U.S. populations regularly since 1999 and provide trends for such health issues as anemia, cardiovascular disease, diabetes, environmental exposures, eye disease, hearing loss, infectious disease, kidney disease, nutrition, and obesity. From 2001 to 2004, an NHANES survey was conducted on balance dysfunction to understand the loss of balance in aging populations and among varying ethnic groups. Balance is an important area and relevant to human health because the loss of balance can lead to falls that are often very injurious, if not fatal.

The NHANES balance survey uses a standardized balance test called the Romberg Test of Standing Balance in Firm and Compliant Support Surfaces. Although the Romberg Test has been shown to have shortcomings, it can describe the ability to balance quite well. The test is based on a simple principle. A subject is asked to stand erect, then to close his or her eyes. If the subject falls, then he or she scores positive on the test. When Friedrich Romberg developed this test in 1846, a positive score would have indicated a lesion of the nervous system. Any test that has the potential result of falling and becoming hurt is not good for the health of the subject, so the test was changed a bit. The original Romberg Test morphed into the very same test that police administer when they suspect someone of being inebriated.

The NHANES balance survey found that no single ethnic group

balanced better than others. Also, no differences were found between the sexes, between smokers and nonsmokers, or between people with hypertension and those without. Oddly, individuals with a high school education scored 40 percent better on the Romberg Test than individuals without the diploma. In addition, people with diabetes mellitus scored 70 percent worse than individuals without the disease and hence were diagnosed with vestibular dysfunction. People with high school diplomas probably read more on the average, and diabetes can cause problems with vision. This information suggests that the last two correlates (a high school diploma and diabetes) might be explained by some vision phenomenon, but establishing causation is another story.

The survey's most important finding is that vestibular dysfunction increases with age, and not just linearly. People between the ages of fifty and fifty-nine were twice as likely to fail the balance test, whereas people over age eighty were twenty-five times more likely to fail. Not surprisingly, the survey also shows a correlation with failing the Romberg Test and falling. So it would seem that we are pretty good at balancing and that some of us can get better by practice. Sadly, though, we lose our acumen for balance as we age, probably as a result of wear and tear on our vestibular system throughout our lifetime.

Speaking of the impact of aging on the senses, hearing is one of the most talked about senses that worsens with age, and it also harbors tremendous variation in human populations. Hearing starts with the ear and the collection of sound waves by the outer ear that are then fed into the middle and inner ears. The middle and inner ears have evolved to detect specific characteristics of sound waves and transmit these to the brain, where they are interpreted and perceived as different sounds.

All waves have two basic properties: height or amplitude and frequency. Observe waves in a bathtub or the ocean; the more force you put behind the wave in the bathtub or the more force the Moon creates to make waves in the ocean, the higher the tide and the bigger the wave, so the greater the height or amplitude. Then note how frequently the waves occur. This characteristic is related to the frequency with which the wave has been generated. The closer the peaks of waves are to each other, the higher the frequency. If the waves are coming in slowly, then they are coming in at a lower frequency. The same principles apply to sound, except instead of water the waves consist of displaced air.

To understand how our ears perceive sound, we need to consider three characteristics of sound: intensity, pitch, and tone. Intensity is related to amplitude, or wave height, and pitch is related to the frequency of the wave. Tone is more complex. Most sources of sound do not make pure sounds, meaning that the sound source will produce waves with multiple frequencies. Tone is related to the purity of the source. For instance, the tone of a band with three instruments is purer when the three instruments have similar pitch or frequency. A band producing sounds with several pitches will sound very different from the band that has a more uniform tone. Notice that I have avoided judging whether pure tone (our three instruments using the same pitch) or a mixed tone (three different pitches) is more pleasing than the other. How the brain interprets the tones tells the individual what is more pleasing.

Although humans can hear over a range of three hertz orders of magnitude (20 Hz–20,000 Hz; see box 1.3), we hear best in the range between 1,000 and 4,000 hertz (fig. 7.2). Sounds with this frequency are relatively high pitched to begin with. Some singing and very high pitched speaking are in this range, but other sounds we process are off the charts.

When Mariah Carey whistle sings at the end of "Emotions," she hits a $G_7$ note at 3,135.96 hertz. She can also hit a low note at 82.41 hertz. To give you some perspective, that annoying test pattern sound you hear every month if you are unlucky enough to be awake when radio stations run the Emergency Alert System test is exactly 1,000 hertz. At the lower end of our range, at 20 hertz, we hear vibrating sounds like the vibration of wind blasting through an open car window when the car is speeding.

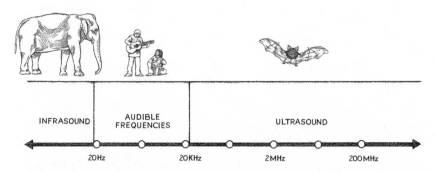

Figure 7.2. Range of sound in hertz, including human audible frequencies.

## BOX 7.2 | PITCH AND MUSICAL SCALES

Although pitch is measured in hertz, it can also be quantified by the notes that musicians use. The seven notes in the diatonic musical scale are do, re, mi, fa, so, la, and ti, starting over again with do. These traditional notes can be correlated with notes musicians use, A through G. Each set of eight notes, do, re, mi, fa, so, la, ti, do (C, D, E, F, G, A, B, C), make up an octave. The highest octave on an eighty-eight-key piano is numbered 8 and the lowest is numbered 0. So, musical scales on a piano can span eight octaves. This makes sense because only one note of octave 8 and three notes of octave 0 are found on the piano. Notes do exist that are off the chart with respect to the eighty-eight-key piano. The frequency of the lowest-used musical note, $B_8$, is only 4.37 hertz, whereas the highest note, $C_0$, is 2,109.89 hertz. Note that this highest note is considerably lower than what Mariah Carey hits when she whistles sings in the song "Emotions." There is an interesting aspect of notes and their frequency that you may have noticed or already know. If the lowest note ($B_8$) is 4.37 and the highest note ($C_0$) is 2,109.89, then the scale cannot be linear (try dividing 2,109 by 4 and you will not get eight octaves). This is because as musicians make sounds that go from octave 0 to octave 1, they double the frequency or hertz output. So, the note $B_7$ is 8.74 hertz, and $B_6$ is 17.46 hertz, and so on. What this means is that even though musicians use these eight octaves, we can also have a higher octave than 8 and a lower octave than 0 (designated with minus signs).

At the upper end, 20,000 hertz, are bizarre rare sounds like high-pitched whistles. Mariah Carey, though impressive, doesn't come close to matching the highest-pitched sound produced by a human and certainly doesn't approach the range a human can traverse with respect to pitch.

Although singers in some cultures can traverse the musical scales, such as the Tuvan throat singers, whose vocalizations are at the low part of the scale of human capacity, the extent of our range of notes is impressive. What then, are the highest and lowest notes ever sung by a human and the greatest range a single human can traverse? The lowest note is seven octaves below zero at $B_{-7}$ and a paltry 0.189 hertz. And although Mariah Carey can hit $G_7$, Georgia Brown, a Brazilian singer, can hit $G_{10}$, or 25,088 hertz, well outside our hearing range. Brown's range exceeds Carey's by four octaves by going from $G_2$ to $G_{10}$, or from 3,135 to 25,088 hertz. But this isn't the broadest range a human can belt out. This distinction belongs to Tim Storms, an American who can span ten octaves from $G_{-5}$ to $G_5$ (0.7973 Hz—807.3 Hz). Storms is the person who hit the lowest note ever recorded ($G_{-5}$). Most humans do not hear such sounds at these extreme ends of the normal range of hearing.

And much like balance, hearing gets more difficult as we age, because as humans age, part of the mechanism with which we detect sound gets worse for wear. As with balance, tiny hairs called cilia are involved. As we age, some of the hairs get broken, and as a result they are nonfunctional. I can attest to this problem with aging. As I get older I tend to not hear my wife as well as when I first met her (or at least that's my story, and I am sticking to it). I like to blame this on her high-pitched voice and not on any lack of attention on my part.

How do our ears perceive the amplitude of the wave? The height of a wave can be translated mechanically into its power. The more powerful the sound, the more of it we detect. So, the mechanical sensors in our ears need to detect not only the peaks of the sound waves but the power the waves generate. They do this by measuring how much of the mechanical parts of the inner ear is displaced. This displacement is what is transmitted to our brain and what our brain interprets as intensity.

The intensity of a sound is measured in decibels, which is a measure of the power per unit of area that hits our inner ears. As the distance from our ears increases, the power of the sound decreases by one over the square of the relative distance moved. So, if a sound is being made 1 meter away, it will have a relative effect of one. But if we move away to 4 meters and make the same sound, its relative effect will deliver power one-sixteenth of the first sound made 1 meter away. If the initial sound was 160 decibels at 1 foot away, then at 4 feet away, it will have a decibel level of one-sixteenth of 160 decibels, or 10 decibels. Unlike frequency, intensity has a threshold of hearing set at zero decibels, which is when the waves being generated by the sound have no power. The sound of a whisper hitting our eardrums is in the range of 20 decibels, and the sound waves hitting our ears from the whisper have the power to displace particles in the air $10^{-4}$ millimeters, or 0.0001 millimeters. A normal conversation at about 60 decibels produces sound waves with the power to displace particles in the air 10 millimeters.

Sporting events bring out the best and sometimes the worst of people, but they certainly are venues for making noise. The loudest recorded roar at an indoor sport was 126 decibels at the Sacramento Kings home basketball court. Although double the decibels of a conversation, this basketball roar is about a million times more powerful than a conversation. The loudest soccer stadium is probably Türk Telekom Arena, home to the Galatasaray football club in Istanbul, Turkey. Soccer fans there have produced roars at 131 decibels, generating a sound with power right at the pain threshold. Türk Telekom Arena is partially enclosed, and so it is pretty astonishing that the fans for a 131-decibel performance have been beaten by an American football stadium that is completely open. The fans at Arrowhead Stadium in Kansas City, Missouri, generated roars at an ear-damaging 142.2 decibels in 2014. This level of sound is well over the threshold of pain and would be much like standing ten feet away from a jet engine without ear mufflers. Rock concerts don't get this loud, even if they turn their amps up to eleven (because as legendary guitarist and Spinal Tap band member Nigel Tufnel points out, eleven is "one louder").

Humans are inundated with sound waves all the time. They are everywhere, and our ears are continually picking them up because our middle and inner ears are built for detecting sound. Our outer ears, those skin and cartilage structures that reside on both sides of our heads, act like funnels for focusing sound into our middle and inner ears, replete with a tubelike channel called the ear canal that focuses the sound waves on the actual biological contraption that collects the sound waves. The structures in our middle ears are every bit as intricate as the balance components of the inner ear. At the end of the ear canal lies a membranous structure called the eardrum. This structure is where the sound waves are collected. As sound waves hit the membranous surface of the eardrum (tympanic membrane) they cause the eardrum to vibrate. Next to the eardrum are the three bones discussed in Chapter 2: the hammer (malleus), anvil (incus), and stirrup (stapes), in that order from outer to inner ear. When the membrane of the eardrum vibrates, it causes movement of the hammer, which is in contact with the back surface of the eardrum. This vibration in turn moves the hammer relative to the anvil, which in its turn moves the stirrup. High-pitched sounds vibrate the eardrum at a higher rate than low-pitched ones. The movement that is transmitted through the hammer, anvil, and stirrup (also known as ossicles) reflects the degree of vibration that affects the eardrum. This whole contraption moves in concert with the vibrations of the eardrum, where the stirrup is connected to the cochlea in the inner ear. The connection of the stirrup to the cochlea is like a piston fitting into a cylinder. Its job is to move the fluid of the cochlea in concert with the information from sound waves that initially entered the ear.

The cochlea is an amazingly convoluted structure. Its three-dimensional structure is like a spiraling snail shell, but even more complex (fig. 7.3). If we could cut across the cochlea to reveal the channels that the spiraling produces, we would see two channels per spiraling arm filled with fluid (the paralymph again). These channels are one on top of the other. In between the two channels lies a duct called the cochlear duct, where the information from the sound waves that initially hit the eardrum is ultimately collected. The duct has two sensitive and movable membranes that lie between the cochlear channels. The duct is also filled with fluid and is connected to an intricate structure called the organ of Corti, which is in turn connected to nerves that run to the brain. The reason there are two channels per spiral is that the paralymph fluid needs to be recirculated. So, if we could run through the cochlea, we would start at the point where the stapes articulates like a piston. We would run through the top canal, and when we reached the very tip of the spiraling cochlea, we would head on our way back out toward where we started in the bottom canal.

The fluid going through the cochlear channels will expand and compress the fluid in the cochlear duct. Such movements of the cochlear duct will impact the intricate organ of Corti. This organ is lined with two kinds of hairs that can bend as the fluid flows in the cochlear canals. One kind of hair is called inner, of which there are about 3,500 cells, and the other is logically called outer, of which there are about 20,000 cells. Bending the inner hairs produces most of the neural response in the neural cell to which the organ of Corti is connected. The bending of the hairs again works much like the mechanosensory bending hairs of the vestibular system and transmits electrical impulses to the brain that are then interpreted by the brain as sound. The outer hairs apparently amplify whatever sound wave signal is coming into the inner ear. The bending inner hairs are kind of like light switches, but they are not uniform in the amount of pressure it takes to flip on the nerve cell. The hair

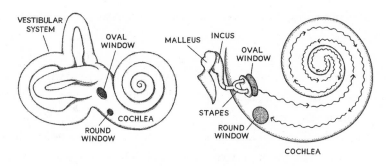

Figure 7.3. The cochlea and its relation to the stapes.

cells near the spiral tip of the cochlea are more susceptible to bending by high-pitched sounds, and the hairs near the base of the cochlea are more susceptible to being bent by low-pitched sounds. Different tones will register on specific hair cells in the organ of Corti. There are a lot of moving parts, but the apparatus works quite well most of the time, as anyone who loves music can attest.

This Rube Goldberg–like apparatus is another wonderful example of how the evolutionary forces at play in forming specialized structures like the one we use for hearing are imperfect. An intelligent engineer would certainly have designed this biological contraption differently.

What would an expert engineer consider good design rules? Amy Smith, an instructor of engineering at the Massachusetts Institute of Technology, has dictated seven simple rules for engineering design. One of Smith's rules is to be frugal and produce the least expensive but most efficient product possible. This rule suggests that getting rid of extra moving parts is wise. The more moving parts there are, the more energy is used. In addition, the more moving parts there are, the more parts there are that could tend to break down and need replacement. Another of Smith's rules is to engineer with transparency so that others understand the product with ease. I don't think anyone who has just read the description of how the middle and inner ears work would say that the design is transparent. Smith's final relevant rule quotes Leonardo da Vinci: "Simplicity is the ultimate sophistication." Her rule is "Do the hard work needed to find the simplest solution."

Evolution simply does not work that way. Evolution works with the raw materials it is given and is somewhat lazy in that respect. It settles on solutions that are the best given the raw materials. In addition, evolution does not allow do-overs or mulligans. In the case of the middle ear, three bones (the hammer, anvil, and stirrup) existed in this area in the common ancestor of mammals and were subsequently affected by the evolutionary process and molded into a single nontransparent, complex, but efficient contraption that in no way approximates the perfect organ that an intelligent designer might engineer.

I have already discussed age as a factor in hearing sounds with high pitches. And what about perfect pitch or the ability to reproduce sounds with perfect frequency or pitch without a frame of reference? Humans vary considerably with respect to this auditory capacity. Using genome survey techniques, researchers have implicated several genes in musical ability, and perfect pitch is one of them. The lack of ability to detect tones, also called tone deafness or amusia, also apparently has a genetic basis. It appears that the variation in these genes is substantial, indicating that, at least tonally, humans constantly hear different things from the same sources (see Chapter 12 for more discussion about the genetics of complex traits involving our senses).

# 8 TOUCHY FEELY

*Touch and How It Is Linked to Other Senses*

"*My own parents were touchy-feely.*" —Ben Stiller, comedian

Touch is the final mechanosensory sense to consider, and it is implemented through the skin (often touted as the largest sense organ in the body). Unlike the other sense organs discussed so far, touch sensors are variable. Some have called the collection of touch receptors associated with the skin a "medley," others a "menagerie." Indeed, unless you're inclined to remember obscure facts, the names of these organs are instantly forgettable. Nowhere as easy to remember as John, Paul, George, and Ringo, Meissner's corpuscles, Merkel cell-neurite complexes, Ruffini endings, and Pacinian corpuscles are the four major somatosensory receptors embedded in the bottom layer, or dermis, of the skin (fig. 8.1). Two other receptor systems are also recognized that relate to nerve endings: anceolate endings and free-nerve endings.

Meissner's corpuscles are specialized on light touch and sensing vibrations hitting the skin. These receptors are the kissing receptors or, more generally, the receptors that send information from the lips to the brain. They are also involved in sensing by touch in the fingers. Merkel cell-neurite complexes are found in the basal epidermis of the skin and in hair follicles. Because of their locations, they can sense low-level vi-

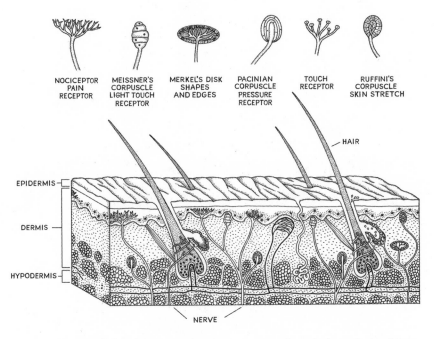

NOCICEPTOR PAIN RECEPTOR

MEISSNER'S CORPUSCLE LIGHT TOUCH RECEPTOR

MERKEL'S DISK SHAPES AND EDGES

PACINIAN CORPUSCLE PRESSURE RECEPTOR

TOUCH RECEPTOR

RUFFINI'S CORPUSCLE SKIN STRETCH

HAIR

EPIDERMIS

DERMIS

HYPODERMIS

NERVE

Figure 8.1. Types of cells that implement our sense of touch through the skin.

brations. Whereas Meissner's corpuscles can detect vibrations with frequency of 10 to 50 hertz, Merkel nerve endings detect vibrations in the range of 5 to 15 hertz. As a result of having what is called a small receptive field, they are most effective in the fingertips, where fine detail is the major focus of touch. Ruffini endings react to skin stretch or distortion. They are slow-response receptors, and they basically tell our fingers where to be when touching. Pacinian corpuscles can distinguish rough and soft objects. They are quick acting and are most sensitive to vibration in the range of 250 hertz, much greater than Merkel or Meissner's bodies. The reader might be asking, then why aren't our brains afire with signals from wearing clothes? The answer is that Pacinian corpuscles react only to sudden stimulation. They quickly forget that they are touching cloth when you put on a shirt and patiently wait for the next sudden stimulation. Free-nerve endings are the tough guys of the mechanosensory group, because they take the touch stimulus and communicate with the brain only after a threshold is reached and that threshold is pain. Lanceolate endings are situated in hairs and follicles and detect movement of the hairs. They cannot detect the direction of the movement of the hairs but are rather good at sensing vibrations at even higher frequencies than the other sensors discussed above (200 Hz–1,000 Hz).

This menagerie of sensory cells, though different in basic structure, all have the same basic floor plan with respect to their connection to the brain. A review of the connections of these receptors from the brain and brain stem outward starts with the connection to the brain stem or the dorsal root ganglia along the spine. From there the cells extend axons called sensory afferents. These axons are the wiring that carries the electrical impulse that signals the brain that something is being touched. The axons then spread out into the endings described above. Information from these touch sensor cells is transmitted to several regions of the brain, depending on the kind of information. For instance, stimulation from the sensors that focus on vibrations is connected to a different region of the somatosensory cortex than sensors that are specialized for surface texture.

As an example of how a sensory receptor is connected to the somatosensory cortex, consider how the brain processes fine touch. When someone touches a small object with the fingertips, Meissner's corpuscles will become distorted by the force of the touch. This will produce a reaction in the sensory cells that in turn produces an action potential. The electrical signal from the action potential travels through the axon of the sensory cell and connects to the spinal cord, where it then travels to the brain. When it reaches the brain, the action potential can take one of several routes to the sensory cortex. Once the signal reaches the sensory cortex, it is processed and linked to the amygdala and the hippocampus as a means to reinforce memory of the sensation. The routes of other action potentials generated by the various mechanosensory cells are similar.

According to David Linden and other neuroscientists, two major neural pathways in the brain are dedicated to touch. The first is the sensory pathway just described and in Chapter 2. Linden points out that this pathway is based in the somatosensory cortex of the brain, where the information from touch ends up. What happens next is that this cortex "figures out the facts, and it uses sequential stages of processing to gradually build up tactile images and perform the recognition of objects." The other pathway is involved in the social and emotional context. The data from the touch and recognition of objects from touching are then interpreted to affect how we behave socially and emotionally in a kind of dual layer processing. And touch can hugely affect our emotional social behavior. In addition, human populations vary greatly when it comes to touch.

Just five years ago, it would have been safe to write that little was known about the human variability of touch. Researchers did know that some humans are exquisitely sensitive to touch. Some people who are on the autism spectrum express what is called touch aversion. They find touch quite unpleasant—not necessarily painful but just unpleasant. This phenomenon more than likely is not because they sense more intensely or are supertouchers but rather because the social aspects of touching are problematic. On the other end of human touch sensitivity, the sense of touch can be diminished in several ways. As with hearing and balance, humans tend slowly to lose the sense of touch with age. Health problems can also result in the loss of the sense of touch. Vitamin B12 deficiency, diabetes, and stroke can result in the loss of touch in certain parts of the body. Inherited syndromes also involve the loss of touch. Riley-Day syndrome, for example, affects sensory nerve cells and produces many symptoms, one of which is decreased sensitivity to touch and other stimuli. It is what is called an autosomal recessive syndrome, meaning first that it is on one of the autosomes in the genome and second that one needs to have two copies of the gene that causes the disorder. This syndrome is found in higher than usual frequencies in people with Ashkenazi Jewish ancestry. Usually a person with the syndrome has inherited abnormal copies of genes from parents who were carriers. These carrier parents are heterozygous (have one normal copy and one abnormal copy) for the gene for the syndrome and don't express the syndrome themselves. The gene that is mutated to cause the syndrome is known, and it is called IKBKAP. This gene makes a protein important in transcription of other genes into messenger RNA. The connection to the disorder is not at all obvious.

Other genetic syndromes exist such as Charcot-Marie-Tooth disorder (CMT). This syndrome is also an autosomal recessive genetic trait. The symptoms are loss of the sense of touch (and loss of ability to sense pain) in extremities of the body such as the hands, feet, and legs. People with the disorder have been known to contract nasty lesions on their extremities as a result of being unable to feel pain there. The disorder also affects balance, not because of problems with the vestibular system but because people who are affected can't judge where their feet and legs are.

James Lupski, a professor and medical doctor, has lived with CMT disorder for more than forty years. Many genetic lesions are thought to be involved in CMT, but Lupski's case was difficult to pin down with the techniques available in 2010. So, he and a team of scientists decided to sequence his genome and the genomes of his family members. By generating the three billion bases of his genome, they hoped to pin down the genetic basis of Lupski's particular kind of CMT. Members of his family don't have the syndrome, and so by cross-referencing the DNA sequence of his family members with his DNA, the team could find the gene responsible for CMT syndrome. In a feat much like finding a needle in haystack, Lupski and colleagues were able to determine that the culprit was a gene called SH3TC2.

Here's how they did it. Remember that each strand of DNA is a long, linear molecule made up of four nucleotide bases (G, A, T, and C) that composes our genes. How the Gs, As, Ts, and Cs are arranged tells our cells to make certain proteins, such as a protein involved in nerve cell structure. Also remember that the genetic code states that different amino acids in a protein are coded for in DNA by triplets of nucleotides. In an illustration of a gene sequence from part of the SH3TC2 gene (fig. 8.2), I separate the sequence at every third base because the DNA triplets code for amino acids in proteins.

The next illustration is the part of the protein for which the DNA codes (fig. 8.3). The letters below the DNA sequence are abbreviations for the twenty amino acids in proteins, and the numbers above the DNA sequence are the positions in the protein numbered from the beginning of the protein. Lupski and colleagues scanned his genome for places where he differed from the reference human sequence (a sequence from someone without CMT) and identified them. These positions are called single nucleotide polymorphisms (SNPs). The researchers had to sift through 3,420,306 SNPs. They rapidly excluded 2,255,103 of the SNPs because they did not lie in regions of known genes. Now they had 1,165,204 SNPs to sort through. So next, they eliminated regions that were in genes but did not code for amino acids (regions like introns). This reduced the number to 18,406 SNPs to search through—better, but still not a trivial job!

gga ctc ctg ata cag gaa ggc cac ttc ttc tgc aga gcc

Figure 8.2. Partial sequence of the SH3TC2 gene.

```
         162 163 164 165 166 167 168 169 170 171 172 173 174

         gga ctc ctg ata cag gaa ggc cac ttc ttc tgc aga gcc

          G   L   L   I   Q   E   G   H   F   F   C   R   A
```

Figure 8.3. DNA and protein sequences of the region of the SH3TC2 gene where mutations will cause Charcot-Marie-Tooth disorder.

As a result of the way the genetic code works, they could eliminate even more SNPs. The genetic code is redundant—for some amino acids, multiple triplets of DNA will code for that amino acid. For instance, CCA, CCG, CCT, and CCC all code for the amino acid proline. If there is a SNP in the third position of the codons that code for proline, it will not code for a different amino acid. No change, no foul, no harm. Such changes are called silent because they don't result in a change in the amino acid their codon codes for. This elimination got the team down to 9,069 SNPs that cause amino acid changes in proteins in Lupski's genome. Next, they used the vast knowledge human geneticists have accumulated over the past century in the Human Gene Mutation Database. This allowed the genomics team to match Lupski's SNPs with positions in known Mendelian inherited diseases and got them down to 54 SNPs in coding regions of genes known to be involved in genetic disorders. Finally, they obtained a list of genes known to be involved in neural disorders, and—lo and behold—by cross listing these, they discovered two SNPs in the neural gene SH3TC2. In fact, they could pin down the exact SNP that was involved in the disorder, and it resides in the 169th codon in the sequence (fig. 8.4).

The change in the DNA sequence in Lupski's genome in these sequences is a C→T (on the coding strand the mutation is G→A), and it

```
                  162 163 164 165 166 167 168 169 170 171 172 173 174

reference         gga ctc ctg ata cag gaa ggc Cac ttc ttc tgc aga gcc

                   G   L   L   I   Q   E   G   H   F   F   C   R   A


Lupski            gga ctc ctg ata cag gaa ggc Tac ttc ttc tgc aga gcc

                   G   L   L   I   Q   E   G   Y   F   F   C   R   A
```

Figure 8.4. DNA and protein sequences showing the location of James Lupski's mutant SH3TC2 gene (bottom).

causes an amino acid shift from histidine (H) to tyrosine (Y). The precise function of SH3TC2 is still unknown, but it is more than likely involved in myelination, or the coating of nerve cells. Myelination acts much like plastic insulation on wires. Without proper myelination, the nerve cells that transmit touch sensations to the brain eventually lose the action potential signal on the way to the brain, and depleted information about touch gets to the brain. This mutation and form of CMT is unique and demonstrated the power of genomics in understanding a neuropathy of the senses in a single individual.

Genetics has been used in other ways to pin down the genes involved in the sense of hearing, but until recently no such studies have been conducted on the sense of touch. For instance, there are more than sixty known heritable hearing syndromes, based on genetic studies, but only a handful of touch syndromes, of which CMT is one. Geneticists use all kinds of tricks to map genes and to look for correlations of genes with phenotypes such as those important for understanding the sense of touch. The common lab model systems have been chosen over the years because they can be manipulated genetically. One of the niceties of these model organisms such as the fruit fly (*Drosophila melanogaster*) and the nematode worm (*Caenorhabditis elegans*) is that they can be bred rapidly and in a controlled manner. Human geneticists do not have this luxury when they ply their trade. The ethical problems with doing human crosses is so obvious that we need not even go there. So, human geneticists use tricks, as in the case of Lupski, where a pedigree is well known for the disorder.

Another technique is to use twin studies. This tried-and-true approach in genetics is important because of the way that traits are inherited. Twin studies exploit the fact that there are two kinds of twins from a genetic perspective. Some twins are monozygotic and arise from a single egg that gets fertilized and then literally splits to make two clones of itself; the two developing embryos are genetically identical, and hence the twins are called identical twins. Other twins arise as a result of two eggs being fertilized at the same time by two different sperm, and because two eggs are involved, they are called dizygotic or fraternal. Dizygotic twins are no more closely related to each other than two siblings born from different births of the same parents. Twins who participate in these kinds of studies are reared together, and that is part of the trick of twin studies. The rearing together ensures that the environments of the two individuals are as similar as possible. Further, identical twins have the same genomes, whereas fraternal twins do not. But when both kinds of twins are reared together, they experience the same environment. Whatever differences there are between them should be caused entirely by genetics, and so by measuring traits of both kinds of twins, researchers can determine the heritability of the traits. Geneticists use an approach that results in what is called the heritability ($h^2$ or h-squared) of the trait. Heritability ranges from 0.0 to 1.0, and traits with heritability close to 1.0 are considered to be nearly completely genetic; those with heritability close to 0.0 are considered to have minimal genetic context.

A pioneering twin study was done in 2012 in Germany using more than three hundred subjects. Of these about two hundred were twins—sixty-six pairs identical and thirty-four fraternal. Researchers obtained heritability measures for two touch mechanosensory traits (fine touch and sensing vibration). In addition, they measured hearing and temperature sensing traits. The researchers were able to show clearly that mechanosensory touch traits have a genetic component. A surprising finding was a strong correlation between touch and hearing traits. By examining test subjects who were severely deaf, they showed that some deaf subjects had terrible tactile acuity, strengthening their contention that good hearing means good touch.

These researchers then turned to known syndromes associated with hearing impairment. One such system they could examine is Usher syndrome because of the concentration in Europe of people with this hearing impairment syndrome. There are three major types of Usher syndrome —USH1, USH2, and USH3—with the severity of the syndrome higher in USH1 and the lowest in USH3. The clinical manifestation is early onset deafness and later-onset retinitis pigmentosa, causing vision problems. Using the same touch tests that they applied to the twins, the researchers examined people who had USH2 (the intermediate type). The subjects had all been genotyped for a specific mutation in a gene called *usherin* (also called *USH2A*) that is known to be involved in the syndrome and codes for a protein known to be involved in the workings of the inner ear. Some subjects had known *USH2A* mutations, and other participants did not. A surprisingly clear-cut result ensued. Those participants with the known *USH2A* mutations also showed poor tactile acuity, and those with genetic changes not in the *USH2A* gene were pretty good at touch. This study identified that the *USH2A* mutation has a common impact on both hearing and touch.

The researchers also studied touch in blind people to determine whether blindness was correlated with tactile acuity. The results demonstrated clearly that there was no correlation. In fact, visually impaired people often have higher tactile acuity than others, suggesting a plasticity in the tactile sense. What we lose in one sense, we can sometimes overcome because of the plasticity of others.

# 9 THE EYES HAVE IT

*The Limits of Sight in Humans*

*"The eye sees only what the mind is prepared to comprehend."*
—Henri Bergson, philosopher

Go to the theater or a concert, and you will most likely see a wide range of people with differing levels of visual acuity. There will be people with and without glasses. Those without glasses may be wearing contact lenses or have had corrective eye surgery. Some of those with eyewear might need very thick lenses, while others need glasses only to read the program. There will be people without glasses reading things far from the stage. There might even be a blind person or two in the crowd. All these are obvious vision differences among humans. Some exist because of accidents or because of environmental exposure or disease, and others are congenital. But there is a lot of other variation in seeing that you wouldn't outwardly recognize at this performance. A small proportion of the men in the audience will not be able to discern between red and green, and some of the women might be seeing shades of red that only they are used to seeing. Some individuals might have tunnel vision (poor peripheral vision), and still others might need to limit the amount of light hitting their eyes and so wear sunglasses. All of this variation is related to how vision works (fig. 9.1).

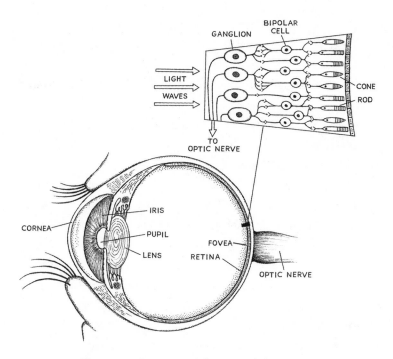

Figure 9.1. Structure of the eye and the retina.

In the eye of a mammal, the path of light to the retina travels along a theoretical line called the visual axis. This is a straight line from the object being observed to the center of the fovea. Along the way to the retina, the light from the object travels through the following layers of compounds and structures. First, it passes through a tear film that covers the eye. This filmy coating protects, lubricates, and keeps the eye surface clear. Next is the cornea, which looks like a clear sheet of tissue but is quite complex, with several specialized layers. The cornea focuses more than half of the light entering the eye. The anterior chamber comes next. This structure is filled with fluid and abuts the following eye part along the visual axis, the iris, which assists in controlling the size of the pupil. The iris is pigmented and is what we refer to when we say a person has deep blue eyes or sexy green eyes or astonishing brown eyes. The next structure along the visual axis, the pupil, is a structure that can expand in diameter to allow more light in to the rest of the eye, or it can contract to restrict the amount of light going along the visual axis. The lens comes next. It is convex and helps to focus light along the visual axis on the retina. Between the lens and the retina is a large structure called the vitreous, through which the light on the visual axis has to pass before hitting the retina. The last structure of the eye in most mammals is the retina.

People with super vision are rarely reported, and these are of such dubious quality that they usually become Internet memes, complete with discussions of the veracity of the claims. Two recent cases are particularly interesting, because—whether they are true or not—they lead us to information about the limits of human vision.

The first super vision story concerns a woman from Germany named Veronica Seider. In the 1970s, Seider was heralded as the human with the best eyesight on the planet because she claimed to be able to see detail on objects more than a mile away. She reportedly could even identify people from that distance, and her vision was thought to be 20/2, perhaps even 20/1. Most humans have 20/20 vision (or 6/6, if you are used to the metric system), which means that they can see objects clearly at 20 feet. If someone has 20/200 vision, this means that an object would need to be 20 feet away for that individual to see detail that a 20/20 person would see at 200 feet. In other words, the 20/200 person has clear vision only one-tenth of most people. Seider's 20/2 vision means that what she could see clearly at 20 feet would have to be viewed at 2 feet by an average person for it to be clear. So, her vision would be about ten to twenty times better than the average human vision, close to if not exceeding the visual acuity of some birds of prey that are at 20/2.

Baseball players who rely heavily on visual acuity when batting need to recognize the speed and spin of the ball well before it reaches the plate to judge where to swing the bat to make contact with the ball. For instance, it has been estimated that misjudging the speed of a pitch by a slight 2.5 miles per hour will result in a swing either 12 inches too soon or 12 inches too late. Different pitches look different to the great hitters. To me, a speeding baseball or a curveball or a forkball or a changeup all look like a blur no matter what is thrown at me (which is why my baseball career ended in high school). But to Wade Boggs or the great Ted Williams, a fastball looks white, a slider looks as if it has a red dot on it, and a curveball appears to have tumbling stripes. Some hitters attain a state of baseball nirvana called precise foveal fixation when hitting. This state is where the hitter is seeing the ball with such precision that judging where it will cross the plate is a snap, and being in this zone apparently does produce a euphoric response. The fovea is an incredibly tiny region of the retina where the most focused, distinct, and detailed vision is received, and it is responsible for foveal fixation. It is best described by the following well-known test. Look at the figures below and focus on the registration symbol (®) for a few seconds, and without losing the focus on that symbol, try to get a sense of how you are seeing the other figures to its right and left.

$$\Upsilon \ \text{H} \ \prod \ \text{M} \ \nabla \ \otimes \ \circledR \ \mathfrak{R} \ \Psi \ \Omega \ \mathfrak{I} \ \text{K} \ \xi \ \prod$$

If you did this correctly, the figures to the right and left of that symbol became a bit blurry. This is because you have focused the registration symbol with the foveal region of your retina, and this is the only thing you are seeing in true 20/20 vision. There is something very special about how the fovea is structured that is essential to understanding vision.

The second Internet meme is the Chinese Cat Boy, whose real name is Nong Youhui. This little boy was reported in 2012 to be able to see clearly in the dark and to have eyes that shine in the dark much like a cat's. Alien conspiracy theorists quickly picked up the story (which is the danger of the Internet), and so the information about Nong is a bit muddled. It is clear, however, that Nong can see quite well in the dark. His shining eyes are the result of a lack of pigment cells in the eye caused by a disorder called ocular albinism. Animals with exceptionally good night vision, such as cats, do have a reflective layer of tissue associated with their retina called a tapetum lucidum that allows the retina to capture more light. The tapetum lucidum reflects light, and so it glows in the dark, but Nong's eyes do not glow because he has a tapetum lucidum. There is no connection genetically or anatomically between ocular albinism and the existence of a tapetum lucidum.

Another eye-related change that animals with night vision have is an increase in the number of cells in the retina. The retina is made up of thousands of small rod- and cone-shaped cells, or rods and cones. Cats have many more rod cells in their retinas, and while Nong's retina was not examined, it is a pretty good bet that he has a significantly larger number of rod cells in his eyes. After hearing these two Internet-based stories I hope you are now curious about the structure of the retina and what these rod and cone cells are and do.

Although the vertebrate eye is complex, with lots of structures used to focus and filter light, the retina is where most of the action occurs with vision, so it is worth describing its structure in detail. To start, remember that the retina is where the action potentials that send the electrical messages to the brain initiate (see Chapter 10 for where these signals end up in the brain).

The retina is literally a field of two kinds of specialized photoreceptor cells, rods and cones, all connected directly to the brain. In fact, many neurologists consider the retina as part of the brain. How these rods and cones are distributed in the retina and what kind of light they are built to detect determines most of what happens with vision. (There is actually a third kind of photoreceptor cell in the retina, the photosensitive retinal ganglion cells [pRGC]. These cells were discovered about a hundred years ago in blind mice. They will react to light even when the rods and cones are missing or incapacitated. The pRGCs are involved in circadian rhythm maintenance and are only peripherally involved in seeing.)

Visual acuity or resolving power is the job of the cone cells, and the fovea, which is the seat of acuity, is packed with cone cells only. The more cone cells there are in the fovea, the more it is functionally aligned with resolution or acuity. This is also where we pick up the best color resolution in the visual system. Assuming that color vision and acuity are connected in some way would be jumping to the wrong conclusion, though. Even though they are both jobs of the cone cells, they are different phenomena.

When there is low light and hence no need for color detection or strong acuity, the rods take over. Not surprisingly, the best low-light vision is away from the fovea and out toward the periphery of the retina, where all the rods reside.

Rod and cone cells are packed closely together in the retina. The ends of these cells that face toward the outer world are packed with proteins that are embedded in the cell membrane and face outward. These specialized photoreceptor proteins are called opsins, and they are structured much like the chemoreceptors introduced in the beginning of this chapter. There are seven transmembrane domains that wind in and out of the cell membrane to anchor the opsin in the rod or cone. As with chemoreceptors, there is a beginning part of the protein lying on the outside of the cell and a little tail of the protein on the inside of the cell. Connected to the interior of the protein, where the seven domains spanning the membrane reside, sits snuggled into the protein a small chromophore molecule called 11-cis-retinal. This chromophore is photoreactive in that when it is struck by a photon of a specific wavelength it will isomerize (change shape but not chemical makeup) and boot itself out of its cozy home in the opsin. In its isomer state the retinal ceases to fit in the opsin's home for it. This eviction event in turn results in the opsin changing conformation and triggering the same kind of G protein reactions we saw with chemoreception when I discussed smell and taste.

Human opsins are a large and diverse set of proteins coded for by genes in the human genome. There are nine major types, but not all are involved in the visual system. The relevant opsins for vision are rhodopsin, red opsin, green opsin, and blue opsin. Important to the story of super color vision in humans is that the green and red opsins reside right next to each other on the X chromosome in the human genome. Blue opsins are located on human chromosome 7, and rhodopsin, the final opsin involved in color vision, is found on chromosome 3.

Light, which is both wavelike and particlelike, consists of photons that can have specific wavelengths. For any photon of a specific wavelength there is an opsin that will react to it. Other opsins will simply sit there and wait until a photon of wavelength that they like hits them. So, for instance, any photons hitting the retina with a wavelength of 557 nanometers (visible light has extremely small wavelength) will hit all kinds of cells in the retina, but only the opsin in cone cells responsible for red color vision will be shaken up. On the other hand, if the photon has a wavelength of 420 nanometers, it will again hit all kinds of rod and cone cells in the retina as well a lot of opsins, but only the blue color opsin in cones will react to it. Oddly, if there is very little light hitting the retina (in other words, it is dark), all of the cone opsins shut down and the rod opsin, a rhodopsin, goes to work. It reacts with photons with wavelengths of about 505 nanometers and interprets the photoreaction and subsequent phototransduction (the signal to the brain) as blue-green color. This is why any night vision that we might have is relatively colorless. Of course, light coming to our eyes exists in a range of wavelengths and not just at 557, 420, or 505 nanometers. So, while photons of wavelength 557 nanometers are what a red-cone opsin reacts to optimally, the opsin will still react, but in a less enthusiastic manner, with light of, say, 550 nanometers. In fact, the red-cone opsin will react with light all the way down to 500 nanometers, but again not as enthusiastically as it would with light at 557 nanometers. The enthusiasm with which the opsin reacts determines the degree with which the G coupled cascade will send messages to the brain and affects how different shades of red or green or blue are detected by the cone cells in the retina.

By looking at the genomes of a lot of people and knowing what kind of color vision they have, researchers have discovered that there are multiple kinds of opsins for both the red-green kind and the blue kind in the cone cells. The different kinds of red-green opsins are called long-wave variants, and there are two major ones: LW (long wave) and MW (medium wave). There is one short wave (SW) for the blue opsin.

The evolution of the LW, MW, and SW genes is a fascinating story. Human variation in seeing colors is best understood by considering the distribution of opsins in organisms in the tree of life. Some bacteria have opsin genes and use these genes as a source of energy from light. The one commonality of opsins across all organisms that have them is that they use a small molecule called retinal as a partner in function. Because light isomerizes retinal, the change in shape of this small molecule has been exploited for many tasks in the tree of life. Plants don't have opsins, nor do some very primitive animals, such as sponges and the pancake-shaped placozoans. But neither of those two early diverging animals nor plants even have nerves, let alone a brain. Cnidarians such as jellyfish, corals, and hydras have light organs and opsins. Ctenophores, the comb jellies, have opsins, too. In some cases, these organisms have a neural net and large numbers of opsins, suggesting that they are detecting broad ranges of light. The box jellyfish, a box-shaped cnidarian cubazoan, has eighteen opsin genes and a complex light-sensing organ that even has a lens!

But it is in vertebrates where the opsin genes really took hold. Although those squishy organisms that immediately precede vertebrates in the tree of life such as sea urchins, sea squirts, and acorn worms have a small repertoire of opsin genes (less than five), vertebrates such as fish, frogs, lizards, and birds all have a much larger number (sometimes more than twenty).

How did the number of opsins jump to twenty? If the animals leading to vertebrates had five or so opsins, then the common ancestor of vertebrates had to have at most five opsins, too. This larger number of opsin genes coincides with a special event in vertebrate evolution, which gives a clue to one way that new genes are generated in the genomes of organisms.

The common ancestor of tunicates (a sea squirt and a hemichordate) and vertebrates again had five opsins. As the tunicates diverged, they maintained the five or fewer gene state. But in the common ancestor of all vertebrates (a different ancestor than the tunicate-vertebrae common ancestor), the entire genome duplicated itself, not once but twice. Whole genome duplications in plants are pretty common (polyploidy abounds in plants), but in animals they are rare. So, this rare double duplication of the genome resulted in the multiplying of the genes in the genomes of vertebrates.

But a funny thing happened on the way to mammals. The number of opsin genes was reset at eight. The platypus, a monotreme and the closest relative to marsupials and mammals, has eleven opsins. So, gradual loss of opsin genes is more than likely how mammals ended up with eight. Alternatively, the platypus might have gained some opsin genes by a process called gene duplication (as opposed to whole genome duplication). This resetting of the number of opsin genes in mammals is not so surprising, given that the ancestral mammal was more than likely nocturnal and had no use for a large repertoire of light-sensing molecules. Eight opsin genes seems to be the sweet spot for mammals, although humans usually have nine in our genomes. So, an even funnier thing happened on the way to primates (box 9.2). The eight opsin genes of the ancestral primate included a rhodopsin, an L/M opsin, and an S opsin, which are involved in color vision. The L/M opsin is a single opsin that has two forms—the L form and the M form.

Of the nine opsins in the human genome, rhodopsin, the S opsin on chromosome 7, and the two opsins on the X chromosome (L and M) are involved in color vision. Humans need all three opsins to be expressed in their cone cells for normal color vision and hence are normally trichromatic. But each opsin can be mutated so that it is nonfunctional, has diminished function, or reacts to a variant wavelength of light. Another possibility is that hybrid opsins can arise as a result of an L opsin (red) gene mixing it up with a green M opsin gene. Because the two genes are similar in their sequences, they can often align themselves across from each other on the X chromosome and a recombination event can

## BOX 9.1 | OUR GENOME

Every human cell in the human body has twenty-three pairs of chromosomes. There are twenty-two pairs of what are called autosomes, numbered from 1 to 22 (the largest chromosome in terms of amount of DNA on it is numbered 1 and the smallest is numbered 22). Autosomes are the collection of chromosomes generally not involved in sex determination. The final or twenty-third pair are the sex chromosomes; in females, there are two of the same kind of chromosome, called X chromosomes. (The X chromosome is one of the larger chromosomes in our genomes with two thousand or so genes.) In males, there is a single X chromosome with its two thousand genes and another smaller chromosome with fewer than one hundred genes called the Y chromosome. Most of the genes on the X chromosome are not found on the Y, so males have only one copy of any gene that resides on the X chromosome.

---

occur, producing two products, one with a red gene front end and a green gene back end and one with a green gene front end and a red gene back end.

All of these possibilities result in a parade of ways that color vision can be deficient in males, because of the X-linked nature of L and M. With two loci on different chromosomes involved and the only combination that gives trichromatic vision (having at least one L, M, and S), several other gene combinations can occur where color vision will be deficient. Males with normal color vision, or trichromatic vision, would be

LM/Y S/S

or

LM/Y S/z,

where the LM/Y represents the opsin genes from your mother's X chromosome and your father's Y chromosome, which lacks an opsin. The z indicates that the opsin is nonfunctional or missing. The S/S and S/z notations indicate the possible opsin combinations from chromosome 7, one from each parent. If you are unlucky enough to be a male with no functioning L, M, and S genes (again represented by a z), genetically you would be zz/Y z/z, and you would be what is called monochromatic and unable to discern color at all. In other words, your cone cells have no opsins at all. You would still have some vision because you would have rhodopsin in your rod cells, but you would lead a very dark and shadowy visual life. This is a condition that actually exists in a large number of individuals on the Micronesian island of Pingelap. Other ways to be a male monochromatic would be

Lz/Y  z/z,
zM/Y  z/z,
zz/Y  S/S, or
zz/Y  S/z.

In two cases, the opsin genes from your mother's X chromosome have only an L (Lz/Y  z/z) or an M opsin (zM/Y  z/z), or in other words, only one opsin instead of the normally linked two, and you would see the world basically in black and white. In the cases of zz/Y  S/S and zz/Y S/z, the cone cells would have only a single kind of opsin (S) in them, and this would produce an extreme black and white world. Females have more chances to make up for missing X chromosomal opsins, and so the frequency of monochromatic effects in females is much lower than in males. Dichromatic males are much more common in many populations. For instance, about one in ten males of northern European descent has one of the following genotypes:

Lz/Y  S/S,
Lz/Y  S/s,
zM/Y  S/S, or
zM/Y  S/s,

and therefore has two opsins in their cone cells. Such individuals are called dichromatic and cannot discern different combinations of colors depending on which opsins are mixed up, most commonly having red-green color blindness.

There are three major dichromatic states: protanopia, deuteranopia, and tritanopia. Protanopes are those individuals who have no functioning M opsins (Lz/Y  S/S or Lz/Y  S/s) and hence no red cone cells. Deuteranopes lack functional L opsin proteins (zM/Y  S/S or zM/Y  S/s) and thus have no green cone cells. Tritanopes are extremely rare; they lack blue cone cells as a result of nonfunctional S opsins (LM/Y  z/z). All of these opsin gene arrangements in humans produce diminished color vision. But there are also some opsin gene arrangements that produce what is called anomalous trichromacy. These are individuals with three opsin types and hence three kinds of cone cells, but one of the opsins is that hybrid L/M gene I introduced earlier. Such individuals are usually males and see in three colors, but the three colors are slightly altered from what a normal trichromat would see.

Note that the color vision system so far describes losses of genes through mutation or even complete deletion, but the opposite can occur. The addition of a functional opsin gene or two to the genomic repertoire is also a possibility. Many animals have added opsin genes to their color perception repertoire to enhance their color detection capacities. And these are not simple additions of opsins outside the color detection range (400 to 700 nm) of humans. That's not to say that many insects and other animals have not added mechanisms for detecting outside of the range of 400 to 700 nanometers.

As noted previously, simple mutations in the opsin genes can change the wavelength of light to which the opsin will maximally respond. If an organism has an opsin that responds best to 560 nanometers, the opsins will maximally respond to light of that wavelength and less so to other wavelengths. Organisms can see color at wavelengths other than the optimal wavelength for their opsin. But if an opsin is added that is specific for a wavelength not covered by the normal opsins, then color vision is accentuated for the wavelength optimal for that added opsin, enriching the color detection of the organism.

For instance, butterflies and crustaceans called stomatopods have taken the game to its limits. Some butterflies are tri- and tetrachromatic but have added extra opsins to fine-tune the color detected. *Colias* or sulphur butterflies have seven color vision opsins, several of which are simply add-ons that act slightly differently. The crustacean stomatopods have an amazing twenty color vision opsins. But they also have six opsins that detect polarized light and two that detect luminescence. The twenty color opsins mean that at least twelve wavelengths are specific for an opsin in these organisms, giving stomatopods color acuity that both dissects and amplifies the usual trichromatic color vision.

Such additions to the human genome in opsin genes would produce what scientists call tetrachromatic individuals. These people would have four kinds of cone cells and would occur mostly in females (since males have only one chance to get the L and M opsins right because they have a single X chromosome). Males are much more likely to have fewer opsin genes than the normal trichromatic state. And females have two chances because they have two Xs, increasing the possibility of being a tetrachromat. Such individuals would have an extra kind of cone cell and would theoretically be able to see more colors than trichromats, who can already discern among millions of colors.

Very few cases of human tetrachromats are documented, and even when a tetrachromat is verified at the genome level, it is difficult to identify the range of color vision these women might have. Having four kinds of cone cells doesn't mean that a brain will interpret what these cone cells detect as different colors, or so the argument goes. Oddly enough, the search for tetrachromatic women usually begins by searching for men with altered color vision. These men are the anomalous trichromats from above. The reasoning is that the mothers of such men gave an X chromosome to their sons that has the anomalous recombinant opsin on it that leads to the fourth kind of cone cell. If the mother also has normal L and M opsin genes on her other X chromosome and two normal S opsins on the two chromosome 7s in her genome, then she will make four cone cells: L, M, and S opsins and the L/M opsin (the recombinant gene which she transmitted to her anomalous trichromatic son). She would potentially be able to have four distinct general wavelengths of light that she could discern with her eyes.

Finding *potential* tetrachromats is not too difficult. These are usually women who self-identify as having extraordinary color vision. There is a simple test to determine if they have the genomic capacity for tetrachromatic color vision, and that is to sequence the opsin gene loci on their X chromosomes and make the call from there. Finding genomic tetrachromats is much more difficult, even though some researchers suggest that up to 2 percent of females on the planet are tetrachromats. There are some compelling web-based stories about women with tetrachromatic vision in the context of art, but these cases are rare. Even when genomic tetrachromats are found, sometimes they simply do not have any better color vision than trichromats. In 2010, Gabi Jordan and colleagues examined a relatively large population using the anomalous trichromat approach and found twenty-four females who were genomic tetrachromats, but after several color vision tests were administered, only one of them could be found to have anything close to tetrachromatic vision.

Kimberley Jameson and colleagues evaluated four human subjects to compare artists' ability and the possible role of tetrachromatic vision that might be associated with that ability. They constructed an experiment in which a comparison is made between two states for each of two variables: an artist-tetrachromat, an artist-trichromat, a nonartist-tetrachromat, and a nonartist trichromat. They then addressed three basic questions with this design by comparing how the four different individuals responded to color vision tests:

- Does the genomic makeup of an individual influence color vision? This test simply compared the artist-tetrachromat and the nonartist-tetrochromat to the artist-trichromat and the nonartist-trichromat. If enhanced color vision was found when there are four opsin genes versus three, then it has a genomic component.
- Does artistic training influence color vision? In this test, the artist-trichromat and artist-tetrochromat were compared with the nonartist-trichromat and the nonartist-tetrochromat.
- Last, do training and genomic makeup influence enhanced color

vision? This test involved comparing the artist-tetrochromat and the nonartist-trichromat with the artist-trichromat and the non-artist tetrochromat.

Only the second test was not significant, indicating that art training is not sufficient to enhance color vision. The other two tests indicate that there is a genomic component and that training and genomic component are synergistic. The obvious caveat to this study is the number of individuals examined, but as a first try at understanding tetrachromatic color vision enhancement it set the bar pretty high.

Modern humans are thus quite variable for color vision. But can we make predictions about our close extinct relatives, such as Neanderthals? Thanks to amazing technology development in genome sequencing, researchers can now look at the genomes of extinct and long-dead specimens. So far, the oldest specimen to yield analyzable DNA is a 450,000-year-old *Homo sapiens* individual found in Spain. More and more long-dead Neanderthal and *H. sapiens* specimens have had their whole genomes sequenced, and so it is possible to look at many of the nuclear-encoded genes in the genome as they existed tens of thousands of years ago. And apparently, the question of whether our archaic relatives saw color like us is an interesting one to boot. The idea is that our modern living styles require enhanced color vision that might have evolved recently after our divergence from archaic humans. In addition, anatomical evidence and the ecological distribution of Neanderthals suggest that they liked dimmer light than modern humans do. John Taylor and Thomas Reimchen examined this question by looking at several Neanderthal genomes and some modern *H. sapiens* fossil genomes. In addition, a third genus *Homo* specimen found in the Denisova Cave in central Asia, called Denisovan, was examined. The authors found no resolvable differences between our current-day opsin genes and the long-dead opsin genes of Neanderthals, Denisovan, and long-dead *H. sapiens*. This result is amazing, given that the genomes tested are all more than thirty thousand years old.

# 10 ACCIDENTS WILL HAPPEN

*Traumatic Brain Injury and the Impact on Our Senses*

*"Maybe when I'd wrecked I had hit my head. Could that be it? Did I have a brain injury? Was I hallucinating? I didn't believe that."* —A. B. Shepherd, author

All our senses require external agents to penetrate the outer shield that our bodies have evolved to keep harmful agents out. It turns out that these protections and mechanisms have also resulted in keeping many signals from the outer world out of our bodies and hence removed from our brains. But the evolutionary process has resulted in a broad diversity of ways that signals from the outer world get collected and transmitted to our brains so that we perceive the outer world more accurately. For example, skin provides protection from dust, dirt, microbes, and other environmental agents that could otherwise harm us, and our skin acts as a first responder to any physical interaction our bodies have with the outer world, such as air pressure or a bump into something. Ears collect and sift through the sound waves that continually bombard us. Our eyes collect, sort through, and transmit the information from light waves to which we are exposed.

All our perception of the outer world, then, starts somewhat in the same place—in cells that act as the first responders to environmental stimuli. Some senses, such as smell and taste, have relatively simple and uniform ways of doing this. Only one first-response mechanism is involved in these two senses, and that is a lock-and-key mechanism. The variety of things we can taste and smell are just variations on a theme for these two senses. Sight is rather like smell and taste in that the first responder is a cell component (a seven-domain membrane spanning protein that is part of a system) very similar to the cell components used in the first response of smell and taste, except the lock-and-key mechanism isn't used. Instead, light hits a cell and pops out an associated molecule (retinal) from an opsin receptor embedded in the retinal cell, and this triggers the further biochemical reactions in the rods and cones of the retina. Touch is unique, and it has coopted several kinds of first responder cells, but again, once they are activated what goes on in these cells is just a variation on a theme involving action potential to the brain. Balance and hearing are also closely related to how the cells in these systems act as first responders. In both of these senses, the bending of small cilia or hairs is critical in starting the response. Even nociception (pain) and sensing temperature act on a first-responder cell whose end job is to transduce a signal (action potential) on to the nervous system.

Once the action potential gets initiated and is passed on from the cell, the nervous system reacts pretty much the same way for all of these senses. The main job of the peripheral nervous system is to get the electrical action potential to the brain. The highways of nerves for the senses are all different, and so the senses reach the brain through a variety of pathways. As illustrated in a few cases, the signals go to very different parts of the brain for the various senses. What happens in the brain is complex, and it is important to understand that the senses aren't simply projected onto the brain by action potential. In fact, a signal from a single odorant will reach many parts of the brain, and the perception of that odorant is actually the combination of action potential going to varied parts of the brain where perception is assembled by thought processes. To tell the story of our senses thus far, I have used the senses of animals and strange human sensing.

Researchers have examined brain function and its impact on the senses in yet another way. The nature of specific brain injuries and how people behave as a result of such trauma or surgical operations can illustrate how clever and plastic our brains are when dealing with sensory input. The approach even has a special name: the clinico-anatomical correlation method. I have already mentioned Wilder Penfield's open-brain surgery experiments. The examples in this chapter will illustrate how the mixture of information and sensory detritus is processed in the brain and becomes perception.

One never knows when an accident is going to happen. Railroad construction foreman Phineas Gage didn't know when he awoke one September morning in 1848 in Vermont that later that day an explosives accident would cause a four-foot-long dynamite tamping rod to blast straight through his skull and out. The result of Gage's accident on his personality is legendary at best and more than likely mythical. When he died, his skull was preserved and deposited in the Francis A. Countway Library of Medicine at Harvard. The archived skull of Gage allows contemporary neuroscientists the opportunity to analyze the connections in the brain that the tamping iron tore through to better understand the impact on the parts of the brain that might have been damaged.

And Franz Breundl, known to science as Mr. B., had no clue when he awoke on a May morning in 1926 in Germany that later that day he would be exposed to overwhelming amounts of carbon monoxide from the smelting apparatus he worked near. The next several months of his life consisted of going in and out of hospitals because of the incredibly bizarre symptoms he suffered as a result of the lack of oxygen to his brain. Mr. B. became a psychological celebrity because his short-term memory had been nearly obliterated by the accident.

Eminent British musician and musicologist Clive Wearing, another unfortunate psychological subject, woke up on a March day in 1986 feeling lethargic. He had no clue that he had contracted a Herpes simplex virus infection. Not long after the original diagnosis, his central nervous system became infected, and some neural tissue was destroyed. Wearing cannot form short-term memories as a result and has experienced amnesia for some of his stored memories.

And on the September day in 1953 that Henry Molaison (also known by the moniker H.M.) went into surgery at Hartford Hospital in Connecticut to have an epileptic condition corrected, he had no indication that he would wake up with no short-term memory. His neurosurgeon, William Beecher Scoville, had performed a bilateral medial tem-

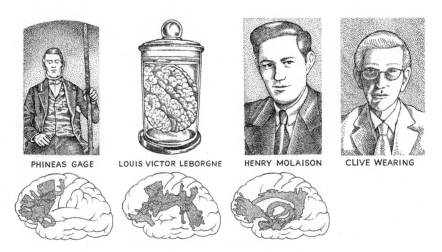

Figure 10.1. Phineas Gage, Louis Victor Leborgne, Henry Molaison, and Clive Wearing, along with the brain tracts of Gage, Leborgne, and Molaison as deduced in de Schotten and colleagues' 2015 study. The only known photo of Leborgne is one of his pickled brain in a jar. After de Schotten et al. (2015).

poral lobe resection, an operation designed to short-circuit the left and right sides of the brain to prevent epileptic seizures.

Hundreds of cases like these four (fig. 10.1) have been documented and studied since brain anatomy became an interesting and important area of inquiry two hundred years ago. Almost two centuries of these cases have serendipitously added to the store of knowledge neuroscience has accumulated about brain structure and function. The French physicians and scientists in the 1800s were among the most skilled anatomists of their time. Their inquiries in the latter 1800s contributed much to what we know today about neuroanatomy. These French researchers also used patients with accidental (or, in some cases, self-inflicted) brain trauma to study the brain.

At a Paris scientific meeting in 1861, Ernest Aubertin, a Parisian physician, presented a paper describing an unfortunate suicide case. In addition to being a physician, Aubertin was also interested in brain function and language and the brain specifically. A Monsieur Cullerier, who had shot himself in the head, was rushed to the hospital where Aubertin was attending. This self-inflicted wound was terrible; part of Cullerier's skull was destroyed and eventually removed by Aubertin, resulting in the brain being exposed. Apparently, Cullerier was conscious during the efforts Aubertin took to save his life, because he could speak. Aubertin then did what Wilder Penfield would do a century later: he placed a surgical spatula on a region of the brain he thought might be involved in language and speech. He asked Cullerier to speak while he applied pressure to this specific region of the brain with his spatula. As Aubertin described in his talk in Paris, Cullerier's speech consisted then of "a word that had been commenced [being] cut in two." When he released the pressure, the capacity to say words returned to Cullerier. Aubertin had manipulated a part of the brain that had something to do with speech. Unfortunately, Cullerier did not survive the day's events.

A few days later, Paul Broca, another French physician who was at the talk, attended to a patient named Louis Victor Leborgne. Leborgne, who was also known as "Tan" (he used the syllable "tan" over and over when he tried to speak), had been in the hospital for most of his adult life as a result of several illnesses, one of which resulted in a loss of speech, known as speech aphasia. He was admitted to Broca's care because he had contracted gangrene of the leg, a condition that Broca could not alleviate, and poor Leborgne died as a result of the infection. Broca, remembering Aubertin's talk, wondered why Leborgne had lost the faculty of speech, so at autopsy, he dissected the brain of his unfortunate deceased patient. During dissection, Broca discovered that Leborgne's brain had a lesion near the posterior end of one of the gyri on the left side of the brain called the inferior frontal gyrus. Two years later, Broca encountered another patient, one Monsieur Lelong, who had had a stroke that resulted in a speech aphasia similar to that experienced by Leborgne. This patient died soon after, and Broca was able to examine Lelong's brain, too. By studying the brains of several more subjects with the same speech aphasia, Broca was able to pin down this area of the brain as one of great significance in the brain for speech, which now carries Broca's name. About fifteen years later, after studying many patients with speech aphasia, the German physician Carl Wernicke was able to associate another region of the brain responsible for language apprehension. This area now carries Wernicke's name (fig. 10.2).

Paul Broca also liked to save brains. Over his career he preserved and archived the brains of 292 men and 140 women for his research. His own brain was preserved and is famously described in Carl Sagan's *Bro-*

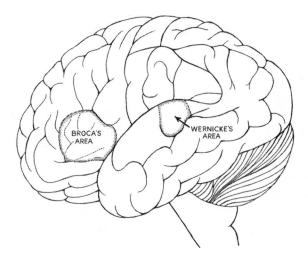

Figure 10.2. Location of Broca's area and Wernicke's area in the brain.

*ca's Brain: Reflections on the Romance of Science.* The preserved brains and the skulls of people with extreme trauma can be examined by modern brain imaging techniques. In a 2015 study by Michel Thiebaut de Schotten and colleagues, three of the very famous brains I discuss above were examined with computed tomography (CT) scans and magnetic resonance imaging (MRI). Both technologies have gained great popularity among neurobiologists over the past decade. The formalin-preserved brain of Leborgne was also visualized using MRI.

CT scans are produced by using a large number of independent X-ray images taken from different angles. The X-ray images are then reconstructed using computer technology to give an overall database that can produce cross-sectional images. The technology can therefore produce three-dimensional reconstructions of the object being scanned from the inside by a mathematical procedure called digital geometry processing. Phineas Gage's skull was examined using this approach, and his long-vanished brain was reconstructed by using the CT scan and information from 129 healthy individuals' real brains. The idea that Thiebaut de Schotten and his colleagues used is that the healthy individual brains and where they sit in the skull would tell them where in Gage's brain the rod had blasted through. Specifically, these researchers could then reconstruct which of Gage's neural pathways were obliterated by the accident.

H.M.'s brain was examined using MRI more than twenty years be-

## BOX 10.1 | MAGNETIC RESONANCE IMAGING (MRI)

Magnetic resonance imaging (MRI) technology takes advantage of how atomic nuclei in molecules and cells placed in a strong magnetic field will absorb and emit radiofrequency energy. These radio waves are emitted because in the nuclei of each atom (such as hydrogen) there are protons that act like tiny magnets. In a normal tissue the nuclei are arranged randomly, but by changing the direction of magnetic fields around the nuclei, the protons in the nuclei will change directions, too, and align to the magnetic field. When the magnetic field is turned off, suddenly the protons return to their original orientation, and the amount and direction of change of the nuclei can be detected by release of energy. This energy has wavelengths on the order of 3,000 hertz up to 300,000,000 hertz (remember that the human hearing range is between 20 hertz and 20,000 hertz) and is the domain of radar waves. The energy released in the guise of radar waves is collected by a set of antennae placed around the object being imaged, and a computer takes the numerical data about the radar emitted and reconstructs the object. More hydrogen atoms (that is, the more water) in a specific kind of tissue or in a specific place in the object will give different wavelengths than an area with fewer hydrogen atoms. So, the image can be quite detailed and give information about an object such as the knee, hip, or brain without having to dissect the tissue or cut into the body.

fore his death, and these images were archived for future use in 1993. In addition, his brain was removed after his death in 2008. It was then that H.M. became known as Henry Molaison, because medical professionals do not use the name of living subjects in their publications or discussions. Molaison's brain was preserved in gelatin and physically sectioned into 2,401 thin slices that were then preserved cryogenically for future research. Each section was subsequently photographically digitized so that the brain could be reconstructed in three dimensions. A web-based atlas of Molaison's brain now exists and can be used as a frame of reference for science. Looking at the images and reconstruction is quite an experience, especially knowing what this person in life contributed to our knowledge of the brain. Take a look at it yourselves and perhaps you will be as impressed with H.M.'s brain as Carl Sagan was with Broca's brain.

The work of de Schotten and his colleagues illustrates that it is possible, with some effort, to reconstruct the brain lesions that afflicted

Gage, Molaison, and Leborgne. In fact, their reconstructions were exquisitely precise. They used twenty-two landmark brain connections and were able to determine the degree of damage to each of these land-marks in the three brains. Although earlier work on Gage's skull indicated extreme damage to the frontal to cortical regions of the brain, the 2015 analysis expanded this gross observation. Because the twenty-two landmarks used are focused on connections of different parts of the brain to others, the researchers could determine whether lesions other than the obvious large frontal and cortical ones existed. Among the several disrupted connections relevant to our examination of the senses, Gage's frontal orbitopolar tract was about 35 percent destroyed. This tract is responsible for connecting the auditory, olfactory, visual, and taste inputs to memory. More than likely Gage's memory of his versions of madeleines was wiped away by the injury he sustained. Molaison's brain connections based on the twenty-two landmarks were the least disrupted of these three famous brains, and this can probably be attributed to the precision of the surgery he underwent that eventually caused his extreme lack of memory. From the surgical records and the MRI done in 1993, it is well known that Dr. Scoville removed several parts of the brain, including the medial temporal-polar cortex, the amygdaloid complex, the entorhinal complex, some of the dentate gyrus, the hippocampus, and other smaller parts of the limbic system. But de Schotten and colleagues' work showed that Molaison's surgery also affected six of the connections that the landmarks could detect. One of these connections is particularly important to the senses and involved the anterior commissure. This neural connection affects the olfactory response and indeed could have affected Molaison, because he had great difficulty discriminating smells after his operation. In two kinds of olfactory tests to identify common items, his accuracy was terrible. He could, however, identify the items by sight, which means that he had not lost the memory of what the things were; rather, as a result of the operation, he was simply terrible at smelling (fig. 10.3). For instance, when asked to identify a clove smell, Molaison answered "fresh woodwork" on the first test and "dead fish washed ashore" on the second.

Leborgne's brain was the most affected by the lesion he sustained early in life. Included in the lesion was not only Broca's area but also several neural tracts that connect it to Wernicke's (fig. 10.1). Other parts

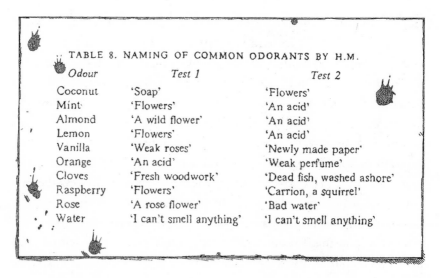

| Odour | Test 1 | Test 2 |
| --- | --- | --- |
| Coconut | 'Soap' | 'Flowers' |
| Mint | 'Flowers' | 'An acid' |
| Almond | 'A wild flower' | 'An acid' |
| Lemon | 'Flowers' | 'An acid' |
| Vanilla | 'Weak roses' | 'Newly made paper' |
| Orange | 'An acid' | 'Weak perfume' |
| Cloves | 'Fresh woodwork' | 'Dead fish, washed ashore' |
| Raspberry | 'Flowers' | 'Carrion, a squirrel' |
| Rose | 'A rose flower' | 'Bad water' |
| Water | 'I can't smell anything' | 'I can't smell anything' |

TABLE 8. NAMING OF COMMON ODORANTS BY H.M.

Figure 10.3. H.M.'s smelling acuity. (Redrawn from Eichenborn et al. [1983].)

of the brain were also affected. Much of the impact of Leborgne's lesion to his behavior and his senses is not known, because Broca first attended to him as he was dying owing to gangrene of the leg. But given the extent of the lesion to his anterior commissure and to nearly every one of the twenty-two landmark connections on the left side of his brain, Leborgne's life post-trauma must have been a nightmare with respect to his senses. To add insult to injury, in presenting their results, de Schotten and colleagues show photos of Gage sitting holding his famous tamping iron (the damage to his head is evident) and Molaison sitting with a wry smile on his face for his portrait, but poor Leborgne's only lasting image is of his brain floating in a jar full of formalin.

So how do neural connections we use from the information our senses gather integrate information from multiple sources to formulate our perception of the outer world? Much of the rest of the story about the senses involves multisensory integration, or crossmodal interactions. These interactions are important for fast, accurate, and sometimes lifesaving interpretation of sensory information (see Chapters 18 and 19).

These famous brains are only a few of the brains that have advanced our understanding of brain injury using the clinico-anatomical correlation method. One reason we have so much information on brain injury and its impact on our senses is that our brains are in the wrong place on our bodies. Evolving to have our brains balanced on our shoulders well away from the center of gravity is a cruel joke of nature. Our brains could not have been better placed for maximal damage when we lose our balance or fall. Actually, it is not so much that our brains were "placed" there, but rather that other things happened that demanded that our brains reside in our heads on our shoulders. Brains were less in the wrong place in our ancestors who did not walk upright and therefore kept their brains closer to a center of gravity nearer the ground. Our species's evolution to upright locomotion dictated that our brains ended up in the worst place possible (other than in our feet, maybe) for keeping them from damage. And humans have figured out some innovative, imaginative, tragic, and downright stupid ways to put brains in situations to be injured. The fact that our brains are prone to injury means that our senses are also prone to disruption, and as we have seen, through examining brains of people with these injuries, we can learn a lot about how the senses and the brain work. We now turn to two of these kinds of injuries that are a scourge of modern life: concussion and battleground head trauma.

On a visit to the Ontario Science Centre in Toronto, I visited the Hockey Helmet Head Hit display in the Human Edge exhibit hall there. This hall is a wonderful example of integrating entertainment and science education in activities for children. The Hockey Helmet Head Hit contraption has a huge red hammer poised to slam into a crash-dummy head with a hockey helmet on it. Pull the hammer back to simulate collisions up to about 20 feet per second (12 miles per hour, or sprinting speed for a hockey player), and let it slam into the head. In addition to offering the thrill of swinging a huge hammer with levers and cranks, the exhibit uses the national pastime of Canada to make a point about how delicate our heads are. The foam head is mounted on another contraption that measures the concussive force delivered by the hammer. And if that weren't enough, the exhibit allows you to choose where the head gets smacked—on the side or in the front. After whacking the crash dummy head a few times at different speeds, one realizes just how dangerous contact sports like hockey are, even with the proper protection. As I played with the contraption, I shuddered at the damage done to the brains of great players like Gordie Howe or Bobby Orr before the helmet era in the National Hockey League. There is no doubt that hits to the head at this speed and in the right place can cause concussive damage to the brain, and indeed, a lot of modern brain research has focused on concussion, how to diagnose it, and what it does to our senses.

About three million adults and adolescents incur brain-related injuries every year while playing sports. Concussions are only one of eight ways to cause traumatic brain injury, or TBI, as a result of impact of the head with a moving or stationary object. Barry Jordan, chief medical officer of the New York State Athletic Commission and medical adviser to the National Football League (NFL), defines concussion as "a complex pathophysiological process that affects the brain and is induced by traumatic biomechanical forces." Sports-related concussion is part of a growing yet underrecognized epidemic in Western countries where contact sports are played, such as American football or, for that matter, Aussie football, hockey, soccer, boxing, rugby, and even simulated martial arts. Many readers will be well aware of the symptoms of a concussion—dizziness, nausea, headache, and loss of memory. Not much is known about threshold of the impact and what concussion is. Even that 12-miles-per-hour hit to the crash-test-dummy head in the Ontario Science Centre might not cause concussion, despite the appearance that any brain inside of a head hit that hard would be damaged.

Because most injuries happen as a result of accidents, the details of many brain injuries are not known. This is why concussion research has focused so heavily on sports injuries. Nearly every concussion in the NFL in the past decade has been documented because of the popularity of American football on television. Researchers do know, by watching films and counting the number of concussions in NFL players and where the players were hit, that with professional players, the biggest damage comes from hits to the side of the head. For children playing football, hits to the top of the head cause the damage. This discrepancy is probably because as one learns more and more about how to play football, one simply does not lower the head readily to hit others, a lesson I learned the hard way as a high school freshman. My football concussion occurred on a kickoff where I wildly threw my body headfirst into the ball carrier. This event taught me to keep my head up and hit and tackle more with my center of gravity, which is with my shoulders and hips square on the target. Unlike the story I told earlier about lacking the athleticism for baseball, my football days ended when I realized I didn't have enough meat on my bones to make hitting with my center of grav-

## BOX 10.2 | CHARACTERIZING TRAUMATIC BRAIN INJURY

There are three ways to study traumatic brain injury (TBI). The first is to study those individuals who come in for treatment of brain injury. One important context for studying TBI is how the injury is incurred, and in many cases brought to the hospital, the precise nature of the injury is not known. The second approach is to use the crash dummy approach like the one in the Ontario Science Centre. The machine used there is not so cleverly named HIT (head impact telemetry). Most research is done this way, but some researchers use model organisms, usually rodents. One device called the FPI (fluid percussion injury) rapidly injects fluid into the skulls of the animals, simulating the role of fluid movement on the impact made during brain injury. Another machine called CCI (controlled cortical impact) uses a device like the one in the novel (and subsequent movie) *No Country for Old Men* that the character Anton Chigurh uses to dispatch his victims. The animal is secured in a stationary position, and a small rodlike apparatus that uses pneumatic pressure extends a piston in the rod that plunges into the skull and into the brain. The depth of the injury and the velocity of the rod can be tightly controlled by the researchers. Another device called the Feeney weight drop controls the dropping of a weight onto the skull of the target rodent. In a variation of the Feeney weight drop called the Marmarou weight drop, a small disk covers the head of the rodent to prevent skull breakage or fracture. The final research tool addresses the exposure of military personnel to explosive devices in the conflicts where they serve. Hence, some of these devices simulate blast injuries by actually setting off a blast with a rodent inside the device.

ity effective. Oh, and by the way, if you are wondering if I ever boxed and got a concussion, yes, I did box—once. My ratio of delivering legal blows to low blows was so bad that the coach asked me to retire my gloves after my first practice. So, my concussions ended when I was in high school, but I can still remember them (at least parts of them), and they were not pleasant.

When a person suffers a concussion, the biomechanical force that disrupts the normal positioning of the head results in rotational movements of the brain inside of the skull. The brain moves in relation to all of the bones in the head, except for structures of the face like the cartilage of the nose. The upper reticular formation of the brain is at the end of the brain stem near where the pons is in the brain stem. It is in a highly conserved part of the brain, is found in all vertebrates, and contains several nerve clusters called nuclei that carry extremely important information to and from the brain. The neural nuclei that course through this part of the brain receive impulses from, and feed to, many other brain regions. This brain architecture makes sense, because after all, this is an ancient part of the brain that mediates very basic physiological and motor control. Electrical impulses from the optic nerve travel first to this formation in the brain to be dispersed to other regions of the brain. Signals from the auditory system do the same, making their first stop in the brain smack in the middle of this cluster of neural tissue. In addition, impulses from the tactile, nociception, and temperature-sensing nerves also pass through this region. What happens is the biomechanical force introduces movement to the brain writ large, which is tethered on the spinal cord. This movement creates torque on the upper reticular formation, that then will incur injury. Often the brain responds by shutting down for a while (loss of consciousness). If that weren't enough, the movement of the brain also results in contact of the brain with the interior of the skull as the brain reverberates from the force of the initial contact. This movement has been described as swirling, and it causes bumping of the brain against protrusions on the inside of the skull that normally do not make articulated contact with the brain.

Perhaps even more dangerous are the effects of brain trauma that occur after the initial injury. These secondary injuries are the result of bruising of the brain and damage to tissues of the brain where contact was made with the interior of the skull and to movement of the cerebrospinal fluid. The damage includes tissue bruising of the brain and deformation of the brain where impact with the inside of the skull occurs. Where deformation occurs, the cells there are prone to dying and their loss of function is expected. Blood vessels get sheared and cause the nerve cells they feed to become nonfunctional. The physical damage to glia and axons (two kinds of neural cells in the brain) causes a cessation of neural activity in those cells that are sheared. Although the primary traumatic brain injury event does not cause breakage of the axons, the impact to axonal structure results in extreme stretching of the axons to such limits that the electrochemistry gets messed up. The cells work overtime to overcome the problem and well up and eventually break. Technical advances like CT scans and MRIs cannot detect these kinds

BOX 10.3 | THE ANATOMY OF A SPORTS CONCUSSION

Imagine a soccer ball glued on the tip of the handle end of a golf club and placed inside a basketball where the inside surface of the basketball doesn't touch the outside of the soccer ball. As long as there is no extreme motion of the golf club, the soccer ball inside the basketball makes little or no contact with the inside of the basketball. But if you jar the golf club hard enough, the soccer ball will bounce off of the sides of the inside of the basketball. I want to say it bounces like a pinball, but not really, because the brain's movement is dampened by the way it is tethered on the spinal column and our bodies. The front of the brain hits the front of the inside of the skull, makes hard contact sometimes with the inside of the orbital ridges (those bones that encircle our eyes) of the skull, and then bounces off. The back of the brain then strikes the back of the inside of the skull. These are called coup and contrecoup injuries, respectively.

of lesions, so researchers use yet a third technical advance called diffusion tensor imaging (DTI) to visualize shearing of axons. This method is pretty spectacular, because it can map the position of specific neural tracts in the brain. It is a magnetic resonance technique, but unlike MRI, which maps overall activity in a brain area, DTI maps the specific tracts that course through the brain. The technique relies heavily on computer processing and is quite expensive. DTI, however, can detect these breaks in the axons as a result of injuries.

A lot of traumatic brain injury research occurs in a military context. From 2000 to 2011, more than 233,000 TBI cases were reported in American servicemen and women serving in the Middle East. Improvised explosive devices and other blasts caused the overwhelming majority of head injuries. Sadly, heads are incredibly vulnerable to injury with explosions and gunfire. These tragic injuries, added to millions of sports injuries, have made TBI a major source of study and information about how injury affects our sensing the outer world. Obviously, bruising the brain and traumatizing the major throughway of information from the sense collection organs like the eyes, ears, tongue, and nose to the brain will affect lots of neural functions generally and how we perceive the outer world specifically.

Nearly every sense is affected by TBI. It has been known for some time that smell is diminished as a result of concussion and more severe kinds of TBI. Trauma to the nose itself has an obvious impact on the sense of smell. The problems caused by TBI on the olfactory bulbs, the neural tracts from the bulbs to the rest of the brain, and the other parts of the brain involved in interpreting smell like the thalamus and amygdala also impact olfactory loss. Loss of olfactory perception is used right after particularly nasty collisions in sports to assess the possibility of TBI. But the degree to which olfactory loss can be used to diagnose concussion or other brain injury has been controversial. However, two studies conducted in 2015 on domestic TBI patients (one in Australia and one in Canada) suggest that between 50 and 66 percent of patients who are treated for TBI have olfactory dysfunction. Nearly half of these patients are extremely affected in their olfactory acuity. In the United States, servicemen who had suffered TBI in Afghanistan and Iraq as a result of explosions were studied for the impact of their injuries on olfactory acuity. The conclusion was that it was possible to correlate actual visible brain injury with olfactory dysfunction only 35 percent of the time. Part of the problem in pinning down the correlation of olfactory dysfunction and TBI involves the tests given to detect the dysfunction. An oddly named tool called Sniffin' Sticks is one of the more popular tests used, but it might behave differently than, say, the University of Pennsylvania Smell Identification Test, or UPSIT.

The clinico-anatomical correlation method has been used only sparingly on TBI patients in the context of olfaction. But the development of DTI technology may offer an important method for studying TBI and its impact on the neural tracts. Patients who have suffered frontal lobe injuries as a result of TBI often have olfactory and gustatory hallucinations that consist of really bad smells or tastes. These dysfunctions substantiate the well-known connection of these two senses with the frontal lobe of the brain. Taste is probably the least studied sense in the context of brain injury and concussion, but some readers who have had a concussion or who have bumped their head will probably remember a metallic taste in the mouth. This sensation is a taste hallucination called parageusia. The metallic taste most likely isn't caused by a dysfunction of your taste receptors on your tongue or of connections to the brain involved in taste. Instead, the impact is most likely on your olfactory system in the brain, and it reverberates in the taste perceived by your brain. (There is a lesson here that we will return to when we delve into how the senses interact with one another in Chapter 11.) Complete loss of taste (ageusia) as a result of TBI would indicate some dysfunction in the sense of taste itself.

Another poorly studied sense in the context of TBI is touch. It is known, however, that TBI damage to the parietal lobe of the brain impairs the sense of touch. This damage will cause tingling of the skin and other touch-based sensations. The parietal lobe is where the impulses from our tactile organs (the many kinds of touch receptor cells in our skin) are processed.

The impact of TBI on vision has been studied in a cohort of servicemen injured by blasts. While the study shows that blasts result in all kinds of vision impairment, oculomotor dysfunction is quite common. This motor-driven phenomenon involves the movement of the eyes in proper vision and results in problems with focusing and with aspects of reading. Computerized tracking of eye movement by individuals affected by blast-induced TBI is being developed as a diagnostic tool and to assess improvement of the visual system in TBI injuries after therapy. Other symptoms of TBI on vision affect so-called higher-order functions, and these include sensitivity to light, reading deficits, and reaction time to sight events. The impact on reading is important because it suggests that the problem isn't entirely related to motor skills. Some TBI subjects complain of losing their place while reading and of not retaining information from reading. These symptoms suggest problems with integration of the information from the eyes as a higher-order process.

Hearing loss in people who have had concussions is also documented and in some instances is used to diagnose brain damage. In the context of TBI and military personnel, it is not difficult to imagine the impact on the auditory and vestibular systems of being exposed to explosive devices. In one of the first systematic, comprehensive analyses of the impact of explosions on American servicemen, Sarah Theodoroff and her colleagues evaluated more than eight hundred publications for information on the impact of explosions on hearing in military personnel. Their results indicate that hearing loss is indeed an outcome of exposure to explosions. One interesting result is that tinnitus, a persistent sound in the ears when there is actually no source of sound, could not be disentangled from hearing loss in the study subjects.

Tinnitus can be divided into two major kinds of problems. The first is pulsative tinnitus, and this occurs when the heartbeat is amplified and can be heard by the person. All other tinnitus phenomena are classified as nonpulsative tinnitus. There are many causes of both kinds of tinnitus, and military personnel exposed to explosive blasts are affected by all of them. These include direct trauma to the inner ear—temporal bone fracture, labyrinthine concussion, disruption of the ossicular chain (hammer, anvil, and stirrup), barotrauma (change in air pressure), and noise trauma. In addition, trauma to the neck and the nervous system (such as the auditory nerve leading to the brain or the areas of the brain involved in processing auditory signals) will also result in tinnitus. As these examples of trauma to the brain show, injury can have a profound impact on our senses. Healthy brains lead to proper sensory perception, but there are other ways to injure or physically alter the brain that will also lead to sensory dysfunction.

## 11 MODERN LIFE, STROKES, AND THE SENSES

*The Impact of Strokes and Other Brain Damage*
*on Sensory Capacity*

*"I had a stroke in December of '99, and it affected my left side—*
*my fingering side."* —Johnny Gimble, Western swing fiddler

What specific kinds of physical damage can occur to the auditory and vestibular systems? When intense sound or air pressure from an explosion comes into the inner ear from the outer ear, it passes across the tympanic membrane, or eardrum. This structure is a thin sheath of skin separating the middle ear from the inner ear and is the first structure that collects sound waves from the outer world as the sound waves hit it and cause it to vibrate. Then the whole contraption of the inner ear starts cranking. Sometimes the intense pressure or strength of the sound waves can cause the thin sheath of skin to become perforated or tear. The effect on the eardrum is a lot like the effect on a snare drum that has been perforated by an overenthusiastic drummer (the late Keith Moon of the Who, who destroyed hundreds of drum kits, comes to mind): the vibrations no longer are true to sound. Another effect is on the stereocilia that line the inner ear and vibrate when affected by sound waves.

The bending of these tiny hairs triggers reactions inside the auditory neural cells, and the signals that these cells send to the brain are regulated via a neural loop in which feedback from the brain adjusts the information coming into the brain. Just like a guitar too close to an amp, if the brain does not adjust the loop, feedback will occur and unwanted sounds will be amplified. The feedback is caused by self-oscillation of stereocilia, and such oscillations in the inner ear can result in tinnitus (Chapter 10). Explosions can break the hairs and kill the stereocilia cells, another way that the feedback can be disrupted in the inner ear and tinnitus can be produced. The receptor cells themselves can be damaged by extreme sound or pressure. Such damaged cells are specific for certain wavelengths and cannot regenerate, so if they are damaged or killed there is no recovering, and deafness to certain wavelengths of sound will ensue. In addition, the lack of function of these receptor cells disregulates the information coming into them, and this will cause tinnitus in addition to the lack of hearing at specific wavelengths. The sounds, in this case, are generated by the disregulation and don't exist except in the head of the beholder.

Hearing can be harmed by insults from other wavelengths in addition to injuries as a result of explosions and concussive hits to the head. We have already discussed the loudest sports stadiums in the world (Chapter 7), but we all experience a lot of ambient noise in our daily lives. This is noise that our ancestors even 500 years ago did not have to confront. Some of the common sounds we hear every day that were nonexistent even 150, 100, or 50 years ago include blenders, coffee grinders, automobiles, televisions, sirens, and ringing phones. Other modern sounds are occupational, such as a plane taking off, a power drill or chainsaw being operated, the sounds in a loud factory, or a subway's brakes screeching. There are recreational exposures to noise, including stock cars or the noise from a Who album as listened to through headphones. And speaking of the Who, incredibly loud concerts like theirs are a major source of entertainment for people. Within this rock group, exposure to their own music resulted in hearing loss of three of their four original members (Pete Townshend, Roger Daltrey, and John Entwistle; the fourth member, Keith Moon, died at the age of thirty-two of excess before it could be determined that he lost his hearing).

If all of these modern sounds we are exposed to lasted only a few seconds, then whether the sound caused traumatic hearing loss, temporary hearing loss, or could be assumed to be safe would depend only on the strength of the sound as measured in decibels. According to Boris

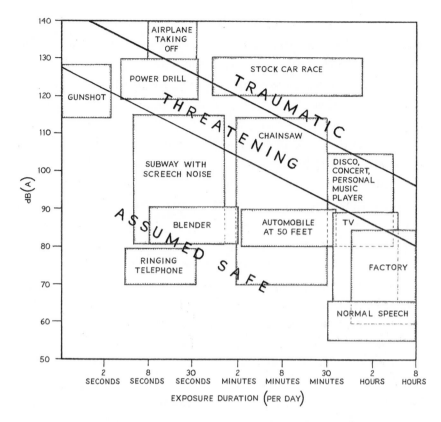

Figure 11.1. Typical sounds and Leq (equivalent sound level).

Gourévitch and his colleagues, though, damage to hearing is caused not only by how loud a sound is but by how long one is exposed to it. To assess the impact of specific modern-day sounds on the sense of hearing one needs to compute what is called a Leq, or an equivalent sound level, for different sources of sound.

This technique for quantifying the sounds we hear in everyday life is based on the decibels the sound makes and the average amount of time each noise is experienced during the day. Blenders and power drills are used in bursts, but power drills have higher Leqs than blenders because they are louder. Two minutes of sound from a blender has nowhere near the Leq level of a power drill sound for eight seconds. The Leq level of the blender run for five minutes is deemed safe, whereas the power drill on for five minutes is deemed threatening and can produce temporary hearing loss. Those noisy stadiums discussed in Chapter 7 are at a Leq level that is deemed traumatic, and if the sound at, for instance, Arrowhead Stadium during a Kansas City Chiefs' game, is prolonged, it could produce permanent hearing loss. And those headphones the baggage handlers wear on the tarmac of any airport are not there for listening to music but rather to muffle the loudest everyday occupational noise that modern humans are exposed to—jet engines.

Noise in the workplace can be annoying at best and dangerous at worst. Most industry standards across the globe dictate that a decibel level of 80 for a full workday is acceptable and "assumed safe." This would be equivalent to listening to cars on a busy freeway for the entire day about fifty feet from the road. Such levels of sound for that period of time should not cause damage to the sterocilia of the inner ear or to the auditory nerve pathways and hence are deemed safe. Indeed, when rats are exposed to sound levels equivalent to those that are at the industry standard, researchers do not see damage to the hairs in the inner ear. But Gourévitch and his colleagues have questioned our certainty that prolonged exposure to high decibel levels of sound are not harmful.

Some research on model animals suggests the opposite. Thirty days' exposure of rats to sound sources at 70 decibels, well below a dangerous sound level, resulted in severe damage to the neural pathways of the primary auditory cortex. These rats could not discriminate between sounds close in frequency that unexposed rats could recognize. The findings indicate that the wiring of the auditory system is affected by noise at 80 decibels for extended periods of time. This rewiring is part of the plasticity that the brain has with respect to its neural connection.

Plasticity of the neural wiring of the brain is a well-known phenomenon, as is evidenced by stroke victims. Stroke causes tissue and nerve cell damage and subsequent loss of sensory, language, or motor capacity that the damaged parts of the brain control. However, because of the brain's plasticity in rewiring neural connections, some stroke victims can regain motor and language capacity through therapy that exploits this ability. Conversely, with exposure to prolonged sound, the brain attempts to cope with the incoming information, and because of the plasticity with which the brain wires itself during challenges from the outer world, it does the best it can, which then causes the damage.

Why not take advantage of plasticity and use prolonged sound in a controlled way to retrain hearing in those people who have diminished capacity at this sense? In 2006, Arnaud Noreña and Jos Eggermont proposed just that: with the knowledge that prolonged sound exposure at high decibel levels can reorganize the cortical connection map for hearing, they hoped to put the principle to some good. They exposed cats to traumatic noise levels that result in hearing loss. After this exposure they separated the cats into groups, treating one with an enriched auditory environment and the other with a quiet environment. This approach is kind of like separating people into the club car and the quiet car on a train. Surprisingly, the group of cats exposed to the enriched environment had lower ranges of hearing loss than the cats in the "quiet car."

Tinnitus, while prevalent in people with TBI (whether it be sports, military, or other injuries), is also found in others in populations who have not been injured. It is thought to affect about 15 percent of the human population and can be seriously disorienting and depressing to those who suffer from it. Researchers have looked for ways of alleviating or curing tinnitus and have concluded that perhaps the best therapy is psychotherapeutic counseling—a sort of mind-over-mind approach called psychoeducation. Since Noreña and Eggermont's sound therapy suggestion in 2006, attempts to use similar approaches have been tried with some success. The best therapies may involve combinations—specifically, noise therapy combined with some other kind of brain stimulation. In rat model systems, targeting the vagus nerve appears to be an important component of this combined therapy. Stimulating this nerve, which is a huge player in neuromodulation, in combination with sound therapy can produce considerable improvements in the level of tinnitus. The vagus nerve stimulation triggers the plasticity of the cortex, and the sound is what the brain rewires on. The stimulation is accomplished by a process called transcranial magnetic stimulation (TMS). Briefly, this treatment involves the focused exposure of parts of the brain to magnetic fields. In this procedure, tones are repeated, interspersed with short pulses of TMS.

Another example of plasticity in the brain concerns strokes. If you have had a stroke, or if you know a relative or friend who has had one, you know that the damage to motor and language skills can be devastating. Strokes are complex injuries to the brain and occur in more than a million Americans every year. They occur because neural cells in the brain are not self-sufficient entities. They need a blood supply to func-

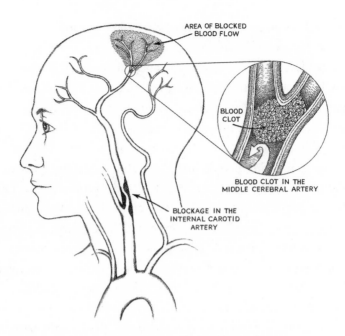

AREA OF BLOCKED
BLOOD FLOW

BLOOD
CLOT

BLOOD CLOT IN THE
MIDDLE CEREBRAL ARTERY

BLOCKAGE IN THE
INTERNAL CAROTID
ARTERY

Figure 11.2. Ischemic stroke and brain damage.

tion, and this is where stroke comes in. There are two kinds of stroke, but both cause death to the neural cells and extreme damage to the neural tissue where the dysfunction of blood supply occurs.

Ischemic stroke occurs when the blood supply to specific parts of the brain is decreased to a level below which the cells can survive (fig. 11.2). Such strokes are localized to areas of the brain where the blood flow is restricted. Blood clots in the brain, embolisms elsewhere in the body, and systemic shock can all cause ischemic stroke because all of these reduce the amount of blood flowing to different parts of the brain. Hemorrhagic stroke is caused when a blood vessel breaks and spills blood into the brain tissue. The burst blood vessels no longer supply blood to the parts of the brain where they are supposed to go, and the lack of blood results in stroke. Ischemic damage can be extreme, but the death of the affected cells is not certain. When the neural cells themselves are killed, they are said to be infarcted. Ischemic regions of the brain have the capacity to recover; infarcted regions do not. More specifically, the lack of blood supply in an ischemic stroke is localized into an area called the ischemic core. This core is usually a small patch of neural real estate around the damaged blood vessels. Leading from the ischemic core is a secondarily damaged area called the ischemic penumbra. This area is also damaged, but not to the same degree as the core.

Recognizing damaged regions of the brain from either kind of stroke requires using brain imaging techniques. The core is diagnosed with a technique called diffusion-weighted magnetic resonance imaging. This approach uses the basic MRI approach but weights the overall image on data that infer diffusion of water. In this way, the tissues that are diffusing blood can be detected, and this is where the ischemic core resides. The penumbra is identified using the perfusion-weighted MRI approach. This MRI technique involves administering a chemical called gandolium to the patient and then detecting the location of the chemical using MRI. Gandolium is a contrasting agent that can be used to identify areas of the brain where the weaker signal of the penumbra resides. A third region, called the benign oligemia, is also identifiable using these brain imaging techniques, but this region after a stroke is, as the name implies, not dangerous because there is little chance of further damage by infarction of the area. The ischemic and hemorrhagic strokes I have discussed so far cause lasting damage to regions of the brain, but there are also transient stroke events called ministrokes. These occur when there is a momentary or brief lapse in storage of blood to a specific part of the brain. The brief stoppage of blood supply to a particular region might result in a temporary and brief loss of the senses, motor skills, or even language. Resupply of blood to the region restores most if not all of the function in the neural real estate that was initially affected. Many of you may have heard of the rather gross term "brain fart" (it's actually in the *Oxford English Dictionary*). The brain fart is simply a momentary and temporary lapse of brain function usually tied to loss of memory. Although it is not a precise medical term, I think that it is quite descriptive in the context of ministroke.

The symptoms of either of these strokes are similar and affect motor skills, sensory processing, and language most acutely. Both tinnitus and hearing loss are the results of some strokes that affect the auditory cortex of the brain. The auditory cortex is localized in the temporal lobe of the brain, so if a stroke affects this region, there is a high probability that hearing will be affected. As you might have guessed, the sense of balance, which uses information from that other sensory organ in your inner ear, can also be affected by stroke. In fact, loss of balance from stroke is as complex as balance itself is and isn't necessarily because the information from the vestibular system gets jumbled. Balance is all about how we sense where our bodies are in space. Our vestibular system of the inner ear does a lot of that calculation for us. But how we sense our muscle movements and muscular tension are also parts of balance, and these are not processed through the vestibular system. Stroke damage to the motor system can result in jerky motion that affects balance. And in fact, stroke can cause loss of sensation on the side of the body opposite the damaged side of the brain, producing a kind of a bodywide neglect. Vision is also involved in balance and is affected heavily by stroke.

Sight is affected by stroke because most strokes occur near regions of the brain dedicated to processing visual information. Decreased vision and double vision are two major visual symptoms of a stroke. Decreased vision is the result of the stroke damaging the optic nervous system and has a high probability of occurring because of the long traverse of the optic nerves from the eyes to the back of the brain to the occipital lobe and on to the occipital cortex. The optic nerve system emanating from the eyes crosses at a point in the midbrain called the optic chiasma. If the region affected by stroke also affects the optic nerves before the chiasma, then any injury to the right (or left) side of the nerves would result in loss of vision processed from the right (or left) eye. On the other hand, if damage occurs after the chiasma (that is, toward the back of the brain past the chiasma), then the pattern is reversed. Damage to the right optic nerve would result in the lack of transmission of information from the left eye, and vice versa. Field of vision is severely affected in all of these cases. Damage to the optic nerves is not the only cause of loss of vision with stroke. The damage can also occur to motor regions of the brain, and double vision is a good example of this. The motor regions control muscles of the legs, the arms, and even the eyes. Double vision is caused by damage to the motor nerves that control eye movement. Because the muscles cannot tell the eyes to align themselves for stereovision, a cross-eyed result occurs and double vision ensues.

So far, I have discussed damage to the visual system that involves upsetting the transport of the primary visual information to the brain. Other damage can occur when nerves involved in the higher-order interpretation of visual information are damaged. A whole set of problems can arise in this higher-order processing category. We'll talk about split brains in Chapter 12, but for now let's discuss four major problems involved in how we react to visual cues and how we read and write using vision. We use our visual system intricately to read and write, and the lack of both the ability to read (alexia) and the ability to write (agraphia) are sometimes the result of stroke. Although localizing the regions of the brain damaged in agraphic and alexic people as a result of stroke have helped localize the regions involved in reading and writing, recent evidence suggests that both of these functions use a surprisingly large number of the brain's regions. Consequently, MRI methods are being used to replace the clinico-anatomical correlation method for localizing brain regions involved in these two brain tasks.

Another result of stroke on visual higher-order processes is a phenomenon called neglect. Although individuals with neglect can see the entire visual field, their brains simply do not process that things are there. Neglect usually is a brain side phenomenon where stroke damage on the left side of the brain will result in the neglect of things in the right visual field. A related problem to neglect is agnosia, where the visual signals are processed all the way through higher processes but the afflicted person cannot recognize people or things he or she sees. The connection of the sufferer's visual system to the parts of the brain where the visual information is interpreted is disrupted, and hence people afflicted in these brain regions are incapable of finishing the visual process of recognizing objects and people. Oddly enough, the perception of color does not seem to be badly affected by stroke. In fact, color is used in rehabilitation of some stroke victims who are alexic. One of the problems that alexia creates is the inability to recognize margins when reading. Colors are often placed at the beginning of lines of type to help recovering stroke victims figure out where margins and new lines are.

Smell and taste of stroke victims are also altered as a result of damage to regions of the brain where the information from taste and olfactory receptors are processed, but less is known about the impact on these two senses. In fact, the loss of olfaction and taste are not considered classic symptoms of stroke. Again, as in vision, both of these senses send impulses to the brain and are delivered to the brain along two sets of nerves specific for the two senses. Stroke can damage the nerves running to the brain and also can damage the higher processing of smells and tastes, or in other words mess up recognizing things as a result of smelling a madeleine. Losing taste might at first glance look like a sense that a stroke victim could dispense with. But loss of gustatory capacity has a huge impact on diet, and because eating has such a social context in modern life, loss of taste can have an impact on family-related activities around which meals are centered. In addition, stroke victims lose a lot of weight as a result of not enjoying food. Finally, whatever taste a stroke victim does perceive is best described as foul and unappetizing, sort of like the metallic taste that TBI victims experience when tasting foods. To compensate, stroke victims have been known to salt or sweeten their food heavily to mask the foul taste generated by damage to the brain. Increased salt and sugar intake, though, is not a particularly good strategy to stay healthy after a stroke.

Stroke and physical injury are not the only ways a brain's real estate can be altered. In fetal development, anomalies can occur that range from anencephaly (lack of development of the entire brain) to spina bifida (incomplete closure of the backbone and membrane around the spinal column). Looking at the impact of these other kinds of brain structure problems can be quite illuminating with respect to our senses.

## 12 FULL/HALF/SPLIT BRAINS

*People with Unique Brains*

*"What it comes down to is that modern society discriminates against the right hemisphere."* —Roger Wolcott Sperry, neurobiologist

Can one survive without a full brain? I have discussed instances where a part of the brain has been removed surgically to stave off epileptic fits, and it does not result in complete loss of neural function. The removal of part of the brain improves the health of the patient and allows the brain to function in some cases. But removal of parts of the brain can also be disastrous. Henry Molaison (H.M.) comes to mind immediately; in that instance, the removal of the inner part of his brain to stop epileptic fits resulted in the loss of his short-term memory and some sensory perception. Although anencephaly (the lack of development of the brain) will be fatal to a fetus suffering from the syndrome, there are less drastic fetal brain development syndromes. Hemimegalencephaly results in a fetus with disproportionate development of the two hemispheres of the brain. In extreme cases, one hemisphere can be greatly reduced so that the person with this syndrome literally has half a brain. But people with this syndrome can live relatively normal lives.

Removal of one of the hemispheres of the brain in an operation called a hemispherectomy is also sometimes a necessary step to stop

## BOX 12.1 | HEMISPHERECTOMIES

This operation is an extreme version of the one performed on Henry Molaison. Of course, doctors would not do this operation if it didn't work, and it is usually performed on children younger than two years who do not respond to repeated drug treatment for epilepsy. Children of this age are preferred for this operation because of the plasticity with which the brain develops in early childhood. Massive disruptions of the wiring of the brain by removal of half of it can be compensated for by neural rewiring as the child develops. Such children lead very normal if not exceptional lives. Either the right or left hemisphere of the brain can be targeted. Some hemispherectomies—ones called anatomical—involve complete removal of the hemisphere. An alternative to this drastic operation is a functional hemispherectomy, where specific parts of the brain such as the temporal lobe are removed and the corpus callosum is severed. The corpus callosum plays a big role in split-brain phenomena. Accidental and operative disruptions of the two hemispheres offer researchers and clinicians a good approach to the clinico-anatomical correlation method to examine sidedness of the brain.

extreme cases of epilepsy. Case studies and long-term follow-up research of children who underwent hemispherectomies and analyses of individuals born lacking a hemisphere of the brain indicate that the visual system is impacted severely. Ahsan Moosa and his colleagues conducted a follow-up study of a large cohort of children with hemispherectomies (box 12.1). The average time from the operation in their study was about six years, so the children had a considerable time for the plasticity of their brains to kick in. The study concluded that vision, though affected by the initial operation, was not significantly affected in 75 percent of the kids at follow-up. This result suggests that there is a considerable degree of corrective plasticity of the visual system in the brain. Several case studies support this result; researchers observe that regions of the remaining half of the brain in hemispherectomies and hemimegalencephalics that are dedicated to vision enlarge with age. Other senses like olfaction and balance are also affected, but as with stroke, these other senses are not focused on in any detail. Speech and reading are also affected by hemispherectomies and hemimegalencephaly (fig. 12.1).

Injuries to the brain (from stroke, surgery, or accident) reveal how

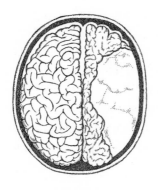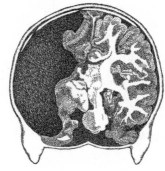

Figure 12.1. *Left,* image of a brain from the top of the head down of a person missing the left hemisphere; the white area is nonbrain tissue and fluid. *Right,* image of a brain from the rear of the skull of a person with a hemispherectomy.

each side of the brain affects function on the opposite side of the body. Understanding some of the nuances of this sidedness is evident in the impact of stroke to vision and to motor skills. If the left side of the brain is damaged, function of the right side is affected. For instance, if we go back to the modern analysis of Leborgne's brain, all of the lesions in brain connections that caused his problems were localized to the left side. Massive damage to the left side of his brain left him using a single word ("tan") when he tried to speak, a function of the right side of the brain.

Anomalies in speech are collectively called aphasias. A transcription from a video recording of an older gentleman, Jack, who suffered damage to Broca's area is poignant, because it is obvious that Jack really knows what he wants to say but simply has to struggle to get the words out, whether they address the question or not. Note the lack of fluid speech and struggle with words that is characteristic of individuals with this kind of aphasia.

INTERVIEWER: What did you do about the ache?
JACK: Uh Home. (pause), Uh Doctor. (pause), And legs. (pause), Walking. (pause), No good.

People who have aphasia caused by damage to Wernicke's area (Chapter 11), which is usually located on the left side of the brain, can speak fluently but cannot string the right words together. Unlike Broca's aphasia, where fluency is nearly obliterated, Wernicke's aphasia results in the inability to recognize what is coming out of the mouth. Here is an example of a gentleman named Byron, who has Wernicke's aphasia.

INTERVIEWER: Hi Byron. How are you?
BYRON: I'm happy. Are you pretty? You look good.
INTERVIEWER: What are you doing today?
BYRON: We stayed with the water over here at the moment and talked with the people for them over there, they are diving for them at the moment.

Psychologists call Byron's response "word salad," because it is a mish mash of words, usually not addressing the original query or a coherent idea. Whereas Jack could intellectually address the interviewer's question, Byron cannot. There is a reason why two older men are given here as an example. Women tend to be able to lateralize, or use, both sides of their brain for language better than men, and there are fewer examples of women showing these aphasias. Although the phenomena I discuss here may seem to have little to do with the senses, it is quite the opposite. Language, speech, and writing are a kind of synthesis of the senses, and higher-level functioning of the brain, which includes speaking and reading, are complex processes involving the senses. Processing the signals of vision, smell, taste, and other senses are just as complex as language.

Broca and Wernicke pioneered brain region studies in the late nineteenth century that culminated in the work of Roger Wolcott Sperry, who received the Nobel Prize in 1981, nearly twenty years after his seminal work on brain sidedness was accomplished in the 1960s. Although Broca, Wernicke, and others recognized the localization of functions to certain regions of the brain, Sperry was able to conceptualize the different functions of the right and left hemispheres. Sperry and his younger colleague Michael Gazzaniga worked with epileptic patients who required surgery because they did not respond to drug treatment. Split-brain individuals endure intentional surgery to sever communication between the left and right sides of the brain. The idea was that epilepsy in some cases is caused by hyperconnected communication between the brain's left and right sides. When the corpus callosum (the region of the brain connecting the right and left hemispheres) is surgically split, the neural connection from one side of the brain to the other is severed. Because the neural circuitry between the left and right hemisphere is disrupted, the epileptic fits cease. Although this procedure can stop epileptic fits, it also results in some strange effects in patients.

The left and right sides of the brain in general need to communicate with each other to properly interpret information from the outer world. The exceptions to this rule have been noted, however, for hemispherectomies and hemimegalencephalics. One surprising result of split brains was Sperry's discovery that the two hemispheres of the brain had different subtasks that are often combined to result in a complex brain function. Interpreting the information from the outer world is a highly complex brain function. After split-brain surgery, the two sides of the brain carry on with their neural tasks, such as gathering visual, olfactory, auditory and other information. But now the two halves do not communicate, so the left half doesn't know what the right half is experiencing, and vice versa. Because the behavior of the split-brain patients is so distinct, Sperry was able to establish several important rules about left- and right-brain function. First, it's really not a left versus a right brain but a dominant brain versus its partner, a nondominant brain. In most humans, the dominant side is the left side. Next, Sperry pinned down that the dominant side (almost always the left side) focuses on and solves analytical and verbally based tasks such as language. The nondominant side (usually the right side) has been thought to be dedicated to emotional and several nonverbal functions. Functions such as creativity have also been attributed to this side of the brain, but these attributes are more than likely not hemisphere-centric.

Another consideration is the context of maleness and femaleness. The general thinking has been that the dominant side of the brain was the female side and nondominant the male. The reasoning has been that women are more verbal than men but men are more spatially oriented. But recent studies of male and female brains have proven this line of thinking wrong. The real difference in male and female brains on average results from how the hemispheres of the brain are wired. Madhura Ingalhalikar and her colleagues looked at the brains of nearly a thousand youths, split pretty evenly between males and females. By using the diffusion tensor imaging method, they were able to map the neural connections of these developing brains. Their results indicate that males tend to have more connections in their brain within hemispheres, whereas females tend to have more connections between hemispheres for the cerebrum.

This result probably means that female cerebral halves are cross-talking more to each other than male cerebral halves. There is a telling difference, however, in the connections between the cerebellum between males and females. In males, there is more cross-wiring from the left cerebellum to the right than in women. Remember that the cerebrum harbors higher-order functions and the cerebellum in general controls muscle movement and coordination. Some researchers think that this is the basis for the general observation that males on average live in a more motor-skill-based world than females. On the other hand, as the reasoning goes, women live in a more intuitive, communication-based world. Although gross overgeneralizations should more often than not be ignored, the difference in wiring is intriguing.

With respect to the senses, the dominant side of the brain is better at expressing what the brain perceives. Verbalization is one of the human brain's favorite things to do, as is evident from our propensity to speak about everything and anything. The nondominant side of the brain is more adept at making sense of or analyzing information. If the information churns up some emotional feelings, that is accomplished in the nondominant side. To conclude that our brains have this dual nature is a major finding about our human condition, and Sperry's contribution was indeed worthy of the Nobel Prize. His student Michael Gazzaniga, perhaps also worthy of a Nobel for carrying the original work to wonderful logical extremes, has focused nearly fifty years of his career on split brains of individuals who have undergone such operations. Gazzaniga throughout his career has exploited an ingenious way of examining how the left and right brains communicate with each other using visual input in split-brain patients.

Split-brain experiments are very logical in their design. Visually, the left eye collects information and sends it to the right side of the brain. Likewise, the right eye collects information and sends it to be processed by the left side of the brain. Next, we need to recall that our brains have evolved to have right-brain-specific and left-brain-specific functions. If the right eye sees nothing, then the split-brain person will verbalize that he or she sees nothing even if the left eye is viewing Andy Warhol's Campbell's soup cans. Strangely though, if asked to draw what he or she sees, the split-brain person will try to reproduce the soup cans. This is because the right eye transmits the "nothing" image to the left brain, which then tries to verbalize what the eyes have just seen—nothing. But the left eye transmits the image of the Campbell's soup can to the right brain, which can interpret it mechanically as a drawing. On the other hand, if the Warhol soup can is flashed to the right eye, the split-brain patient will answer something like, "I see a Warhol." By setting up a system whereby the left eyes and right eyes of split-brain patients view different pictures or items and then asking questions about what the eye sees, researchers can discover amazing intricacies about how our brains deal with visual junk.

One of the more famous split-brain experiments concerns flashing the word "face" to the left eye and the word "smile" to the right (fig. 12.2). The split-brain patient is then asked to describe what he or she saw through drawing and then asked to explain the drawing verbally. In one case a patient was asked to draw with his right hand (which is controlled by the left brain) what had been seen. He drew a smiling face. Sounds about right. But when asked to describe in words why the smiling face had been drawn, the patient, who can't integrate both words verbally, makes up a stunningly clever explanation. To justify drawing the smiling face, the patient answers that he drew a face, and a smiling face is more pleasant than a frowning one. And he concluded by saying, "Who wants a frowning face around?" Strange, but completely explainable by right brain–left brain dynamics. The left brain, having limited information, makes up a logically pleasing story to compensate not only for the lack of information but also for the haunting need to draw the face with a smile because of the lingering specter of seeing the word "smile" with the right eye. Psychologists call this phenomenon of justifying and unifying what is seen from both eyes by the split-brain patient a "unified sense of self and mental life."

One of the more interesting experiments Gazzaniga and his colleagues performed with split-brain patients concerns the recognition of self. He asked, "Severing the corpus callosum in humans has raised a fundamental question about the nature of the self: does each disconnected half brain have its own sense of self?" A basic understanding of how we perceive the outer world would need some partial answer to this question. By using a face-morphing technique, David Turk and his colleagues made the question "Is it Mike or me?" A split-brain patient's

Figure 12.2. Split-brain experimental design.

(JW) face was computer morphed by increments of a tenth with the face of a long-time associate who just happened to be Gazzaniga (MG). In other words, a facial spectrum going from left to right was created with JW's face at the left end and MG's at the right. The eight faces in between looked 90 percent like JW and 10 percent like MG, then 80 percent like JW and 20 percent MG, and so on. In other words, if JW was not a split-brain person, then the photos would look, from left to right, 100 percent self, 90 percent self, 80 percent self, and so on.

Next, Turk and colleagues used the classic approach of exposing the morphed images to the left brain (via the right eye) and to the right brain (via the left eye). At each exposure, JW answered the question, "Is it me?" or "Is it Mike?" The outcome is that the left hemisphere rapidly detects partial images of the self, and kind of linearly. The right brain, on the other hand, can recognize self only with a nearly full picture of self. More precisely, the morphed face needs to have at least 80 percent of self in it to be recognized as self. Because the left or dominant hemisphere recognizes self even with a very little of self there, it suggests a stronger role for the left hemisphere in what Gazzaniga calls "retrieval of self knowledge." But the experiments do not imply that each half of the brain has an individual sense of self; rather, they show that sense of self comes from specialized functions of both hemispheres together.

By far the most bizarre case of the human brain dealing with visual information and turning it into perception is found in the work of V. S. Ramachandran and colleagues. Ramachandran is most famous for his work on synesthesia and phantom limbs, and he uses the example of Capgras syndrome to demonstrate how needy the brain is with respect to creating explanations for unexplainable sensory input and how this is an important part of perception. Individuals with Capgras syndrome will claim that people close to them are impostors. In the case that Ramachandran studied, the male individual claimed that his mother was an impostor. He would be introduced to her and claim something like, "She looks like my mom, but she isn't." The emotional response of this person was measured when he was confronted with his mother, and this analysis indicated that he simply responded neutrally to his mother—every mother's nightmare and more than likely producing overwhelming guilt in the son.

Ramachandran offers the following explanation for this bizarre behavior. The person he examined was an individual who had suffered head trauma. This individual's limbic system, in particular the amygdala, a region deep inside of the brain responsible for emotions, had been damaged. In addition, more than likely the connections of the temporal cortex to the limbic system had been altered. Another important piece of information here is that a region of the temporal lobe called the fusiform gyrus is responsible for processing facial images and for facial recognition and is also connected to the limbic system, just as many other regions of the brain are. So, the individual sees his mother, recognizes her as such, tries to send this knowledge to his amygdala for emotional processing, which is thwarted, and is left with the only logical explanation possible—this isn't my mother because I lack emotion for her. His brain, trying to make sense of some messed up stuff he is seeing and thinking, makes up the impostor story. The clincher for this explanation is that if his mother calls him on the telephone and speaks to him, he recognizes her voice and properly sends this information to his amygdala and has the emotional response his mother adores. No more imposter.

Michael Gazzaniga had this to say about split-brain studies when imaging technology began to explode on the research front: "I have no doubt that the interplay between split-brain research and other methodologies such as neuroimaging will continue to shed light on the human mind and brain." However, the split-brain studies that I have used to introduce the pathways with which our brains use to process sensory information are a thing of the past. Because surgeons performing the operation concluded that the surgeries were not as effective as they should be, they were discontinued. Many of the split-brain patients who resulted from the surgery are now dying of natural causes, and with them is disappearing the ready opportunity to study split-brain phenomena. Hemispherectomies continue for very young children, but because there is so much wonderful rewiring of the brain as these children develop to adulthood, split-brain effects are not pronounced enough for researchers to exploit. Accidents will continue to happen, and these unfortunate events, when they occur to very specific regions of the brain, can produce split-brain phenomena. Researchers have for some time suggested that individuals who are afflicted by a rare birth disorder called agenesis of the corpus callosum (AgCC) might be the saviors of the split-brain paradigm. Only more recently have such cases become important because of the dwindling population of surgically induced split-brain patients. These unfortunate AgCC individuals are born with complete or partial absence of the corpus callosum. Without a corpus callosum, the neural fibers that make up this structure and run latitudinally to connect the brain hemispheres instead develop in a longitudinal pattern within the hemispheres. Partial disorders of the corpus callosum go by different names, but the effect is the same as with the surgeries that cause split-brain phenomena.

There are some distinct similarities of surgical split-brain people and people with AgCC. Although AgCC people have limited connections between the hemispheres of the brain, they seem to have integrated the two hemispheres more than surgically produced split-brain people. The age at which the AgCC patient is examined is also a factor in the connectivity of the two hemispheres, indicating that neural plasticity may in some cases compensate for the lack of connectivity as children with AgCC develop into adolescence and adulthood. One major similarity of AgCC and surgically caused split-brain people is that they are severely impaired in dealing with complex situations. Sadly, patients with AgCC manifest many of the characteristics of autistics. One of the more famous and visible cases of AgCC is connected to the 1988 movie *Rain Man,* in which actor Dustin Hoffman plays an autistic adult. The person on which his character is based was Laurence Kim Peek, who died in 2009, and was considered a megasavant because of his capacity to remember things. Hoffman's portrayal of Peek was a wonderful and complex portrayal of a man with autistic characteristics caught up in the world outside the institution where he had long lived.

In 2013, Pratik Mukherjee and colleagues examined several people with AgCC and determined not only that the corpus callosum was affected by this developmental disorder but also that the cingulate gyrus showed abnormalities. It is well known that this region of the brain is critical for processing information and placing it in an emotional context. Without normal connectivity to this region of the brain, the emotional response to sensory information is lost in the missed connections. This study explained a lot of the behavioral attributes of Peek and others with this unfortunate developmental syndrome. How studies of AgCC people will be incorporated into split-brain research is another story, but such people and their curious brain structures caused by developmental problems might be an important inroad to understanding the nuances of how the brain processes complex sensory information that lead to emotional-, logical-, and perception-based responses to the outer world.

One clue is how AgCC patients react to and interpret proverbs. Proverbs are those catchy little one-sentence statements that need to be interpreted in nonliteral contexts for the point of the sentence to be comprehended, such as "You can't judge a book by its cover." It turns out that AgCC patients fare pretty poorly on the proverb tests compared to peers who have an intact corpus callosum, meaning that there are some basic differences in how AgCC individuals process complex sensory input.

Overall, the human brain is pretty amazing at coping with the signals from the outer world. Humans have developed some astonishing, clever, unique, and sometimes logic-defying neural mechanisms for coping with their sensory world. And these are especially interesting when the senses interact in crossmodal ways. In fact, more than likely few of our sensory experiences are mediated by a single sense.

## 13 TEAM OF RIVALS MEETS THE KLUGE

*Making Sense Out of Crossmodal Stimuli from the Outer World*

*"And the blind man said to the deaf man, 'Do you see what I hear?'"* —Wayne Gerard Trotman, filmmaker

Catchy names for our brains and how they work are aplenty. Two of my favorites are "kluge" and the "team of rivals." I have come up with a few more, which I will not mention here out of embarrassment, but I like these two monikers because they encompass the evolutionary, psychological, and neurological context of our brain. The word "kluge" as a descriptor of the brain comes from neurobiologist Gary Marcus and his book of the same name. Marcus describes our brains as functional Rube Goldberg machines using the German word *kluge,* defined as "an ill-assorted collection of parts assembled to fulfill a particular purpose." "Team of rivals" comes from David Eagleman, also a neurobiologist, from his book *Incognito.* His way of describing our brains plays off the well-known strategy Abraham Lincoln used in assembling his cabinet during the Civil War (as made famous by historian Doris Kearns Goodwin).

On the one hand, the brain has this messy makeup that defies most architectural or engineering logic. Some have likened it to a computer, but this analogy is mostly incorrect because computers have not evolved the way brains have. It is true that computers have changed over time, and it could be argued that they have evolved, but brains have had a much more checkered history than computers. There are no mulligans or do-overs in evolution, so the structures we see in organisms, and especially in brains, are the result of common ancestry. Hence brains are molded by the historical contingency of the evolutionary process. In fact, the best and most innovative solutions in computing are probably those that scrap a good deal of previous work and start with a relatively clean slate—a computer mulligan.

The human brain is a pretty bad example of good engineering, but it works. Once a structure or behavior arises as a product of mutation and is amplified as a result of natural selection, or a structure or behavior arises and is amplified by genetic drift, no matter how klugey, it is retained in a population. And if natural selection acts further to increase the frequency of the klugey solution, a population cannot simply scrap the solution for a less klugey approach. Quite the contrary, natural selection is forced to use the variation that exists, and if the variation is klugey, then chances are that the product of natural selection will be even more of a kluge. There are cases where the phenotype of individuals in a population change significantly and rapidly to give a product of natural selection pretty different from the original variants (by genetic drift or other evolutionary-developmental processes called hopeful monsters), but for the most part, the variants that are there are what natural selection works with. This doesn't mean that things, once klugey, will *always* get klugier. But it does mean that kluginess can beget even more kluginess, and this is more than likely what happened with brains, as exemplified by following the changes throughout the tree of life and especially in the vertebrate part of the tree of life.

Eagleman's team of rivals analogy is also a wonderful descriptor of the neurological context of the brain and how we sense the outer world. According to historian Doris Kearns Goodwin, Abraham Lincoln's strategy for filling his cabinet in the 1860s was to enlist people who he knew would conflict with him and each other and would thereby produce more than conflict for the sake of argument. These rivals worked well together despite their ethical dilemma over slavery, conflicting political views, the threat of secession, and the ravages of the Civil War. The brain takes in all the contradictory information from the outer world gathered by our senses, the stimulation of chemicals from our nervous and hormonal systems, our physiology, and other extraneous and rival information and interprets it to keep us functioning. Conflict and rivalry then become important aspects of how our brains work. If there is no conflict, then there is no problem, and the operation is completed without a hitch or hiccup. For instance, Eagleman uses a car turning a corner as an analogy of something that is not conflicted and, hence, no problem. A steering wheel and the driver who turns that wheel control a car. A car does not complain about what it is doing when turning, and hence there is no conflict. The brain does not work that way. How we turn outer world stimuli into perception can be looked at and analyzed from two directions: bottom up and top down.

The brain is continually exposed to conflict from rival signals and information. The problem is to decipher how we process the information. Psychologists seem to think that we can do this in two general ways. If the information coming into the brain is optical in nature, then it takes a circuitous route in the brain to determine shade, color, shape, and other aspects of what we have seen. Memory and emotions kick in to give an overall perception of what our eyes just saw. Because this way of perceiving starts with data or information and the information is built upon by other neural functions with increasing complexity, psychologists call this a bottom-up approach to perception. The other approach to perception starts with the sensory information triggering memories of and emotions about things we have seen and interacted with in the past. A database, so to speak, exists in the brain that reacts to initial stimuli. Using this experiential reference, our senses then track the information through other parts of the brain to construct a perception. This route is pretty much the opposite of bottom up and is called the top-down approach to perception. The latter sorting of information (top down) to create perception is triggered in the context of memory, emotion, and other higher-order functions of the brain.

The importance of neural connections in how we use the information from our senses and integrate the information from multiple sources to formulate our perception of the outer world is a matter of integration. But is it top down or bottom up? In fact, in many ways it doesn't matter which it is, and perhaps it is even a mixture of the two approaches to perception that acts in our brains. Just realizing that these are two possible processes allows psychologists to formulate testable hypotheses and create clever experiments to test the hypotheses about how perception works. Although this approach focuses on being able to reject hypotheses, in this experimental framework often we cannot reject anything in the context of reality. But what comes out of the experiments in this kind of scientific method is both greater understanding of how the brain processes information and more refined hypotheses of perception.

The neural pathways in the brain that take information from the outer world through the sense organs and travel through the brain are often circuitous. The routes through the brain for each of the big six senses vary, indicating that different parts of the brain are responsible for processing different sensory information. We have already seen how the sensory cortex is involved in processing touch, and although we can map the pathways to their processing points by means of homunculi, there are other parts of the brain where signals from the sense of touch must interact with memory and other higher functions. Perhaps the best-understood sense for the intricate pathways that are involved in interpreting sensation is sight (fig. 13.1). Years of neuroanatomical and psychological work have pinned down pretty precisely the major pathways in the brain for vision.

Vision starts within the eye and the neural impulses that are created by light hitting the rods and cones in the retina. The pathway that the impulses travel from the retina to the brain is decipherable simply by tracing the anatomical structures emanating from the eye. The nerve cells coming from the eye are bundled into rather large neural structures, the optic nerves. The optic nerves from the two eyes cross from the left eye to the right side of the brain and from the right eye to the left side of the brain via the optic chiasma. Just past the optic chiasma the two nerve bundles reach farther into the brain and connect to structures called lateral geniculate nuclei, one on each side of the brain. The two lateral geniculate nuclei then serve as relay stations that send impulses farther to specific regions in the back of the brain, where the optic nerve cells start to radiate. The discovery of the functions of the radiations or streams of neurons is discussed in detail below. The processing of the information doesn't stop with these radiations or streams that reach into the visual cortex, but then loops out to the prefrontal cortex, where the information is placed into working memory so that we can easily access

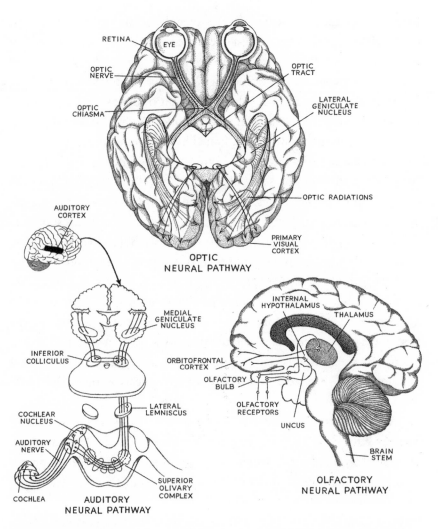

Figure 13.1. Neural pathways for vision (top), hearing (bottom left), and smell (bottom right).

the information for further use. What happens in each of these brain areas has been the subject of much research and has revealed a great deal about how complex the brain really is.

A century ago, using the clinico-anatomical correlation method, several German neuroanatomists pinned down that a specific region of the brain, when injured, would result in a visual anomaly called agnosia. Their observations allowed them to generalize that people with severe injuries in two regions of the brain—the lower region of the temporal lobes and the ventral occipital cortex—were unable to identify objects placed in front of them. More specifically, two gyri (the convex rolls of the human brain)—the lingual gyrus and the fusiform gyrus—were most often the site of brain lesions causing visual agnosia. Since the eye itself is not the source of the cause of visual agnosia, the actual visual information from the outer world is not deterred from entering the brain. Rather, the damage to these specific areas of the brain prevents processing of the information and results in the agnosia. In some rather brutal experiments, two neurobiologists in the 1930s removed chunks of rhesus macaque (*Macaca mulatta*) brains to determine the impact of lost function from these regions. Heinrich Klüver and Paul Bucy conducted the now famous and Hannibal Lecter–like experiments that bear their name: Klüver-Bucy syndrome. Although the removal of a big chunk from one side of the brain can often be overcome by rewiring, Klüver and Bucy did what is called a bilateral removal. In other words, they removed the corresponding chunks from both sides of the brain. In so doing, Klüver and Bucy produced monkeys that were very messed up with respect to vision, which made them very messed up in other behavioral aspects. They were unable to correctly or even slightly recognize images, and this lack of visual capacity severely altered several behaviors dependent on image recognition, such as eating and sex.

Brutal as these experiments might be, and indeed they would probably be condemned today by many animal rights advocates, they did lead to an explication of the neural pathways involved in vision. It is safe to say, though, that without these macaque brain ablation studies, we would have had to rely on the vagaries of the clinico-anatomical correlation method and more than likely would have only a very partial picture of the pathways involved in vision. The most important work from these ablation studies revealed two pathways through which action potentials involved in sight are processed. It turns out that there is an upper (dorsal) pathway and a lower (ventral) pathway of neurons that process visual information. The ventral stream takes care of the "what" of objects in perception, and the dorsal stream takes care of the "where" of objects in perception (box 13.1).

Within each stream neural impulses take well-defined pathways, and hence each part of the brain in these streams does well-defined tasks (fig. 13.2). For example, how does the brain process color?

## BOX 13.1 | WHAT AND WHERE

Heinrich Klüver and Paul Bucy used ablation studies in macaques to show that disruption of the ventral pathway (or in macaque brain anatomy terminology, the occipito-temporal stream) by removal of chunks of the lower temporal lobe resulted in monkeys who cannot discriminate among objects. These were essentially monkeys who probably mistook their mates for a banana. Although these monkeys had trouble with figuring out "what" things were that they saw, they retained spatial visual acuity and could easily judge perspective and distance of objects ("where"). Ablations in the dorsal pathway (the monkey anatomy term is occipito-parietal stream) produced monkeys who could identify objects but had huge difficulty with spatial vision, or the "where" of objects. These results led to the understanding of the "what-and-where" dichotomy of the ventral and dorsal streams of visual information.

Color is part of the "what" that gets processed in the brain in the ventral pathway, so the color-processing part of the brain is in the ventral stream in the temporal lobe. In addition to color, this part of the brain also processes shapes, shades, and textures, except that these other aspects of "what" are processed in their own subareas of the ventral

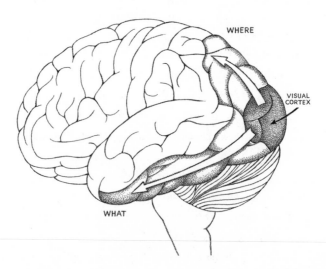

Figure 13.2. The visual "what" and "where" neural pathways in the human brain in the visual cortex.

stream called "V" areas. These ventral processing functions are distinct and relegated to very specific regions of the brain and are thought to be sequentially connected. In fact, the V regions are numbered, and their numbering reflects their place in a spatial hierarchy of each of the two streams.

What about something that is moving? Here we want to know about "where" the thing that is moving can be placed in space. So, this kind of information is "where" information, and as we said earlier, "where" is processed in the dorsal stream visual pathway. Indeed, things like motion, direction, and speed of objects are processed in the dorsal stream and again in unique V regions of the brain. The route through which action potentials move can be straightforward or circuitous depending on the difficulty of pinning down the "what" or "where" of the object. In other words, although researchers have understood the spatial location of the V regions and other regions of the temporal and parietal lobes important in processing visual stimuli, the route the neural impulses take is not only nonlinear but also nondirectional. The neurons in the different areas of the brain where the information is processed have multidirectional function. This means that vision is not necessarily unidirectional from neurons deciphering less complex perceptions to neurons solving more complex perceptions. The system is best described as having massive feedback and feedforward functionality. In addition, some linkages skip hierarchical levels in the ventral and dorsal streams. And to make the situation even more complex, there are potential connections within the individual V regions of the visual pathways that are critical for processing visual stimuli.

How does this neural architecture fit with top-down and bottom-up processing of visual stimuli? It means that both types of processing are possible. Anything with a feedforward neural pathway would be part of a bottom-up process, and anything that might feed back would be part of a top-down process. And it is not hard to visualize neural responses to visual stimuli that bounce around a bit with both feedback and feedforward patterns, making some of the processing of a cohesive visual stimuli a top-down process and other parts of it a bottom-up process.

The neural pathways for the big five senses are fairly well known, and the general theme of the route through which the initial action potentials from the external sense organs travel follows the overlying thematic tenet of sensory pathways architecture I call "It's complicated." Smell and taste receptors are triggered chemically. The action potentials produced by chemoreception in the nose have a relatively short distance to travel, because the olfactory bulbs that are the first stop for the impulses lie almost directly above the receptors for smell. The impulses take this somewhat direct route to the brain and pass through the olfactory bulbs, where primary processing is accomplished, and on to the primary olfactory pathway (see fig. 13.1). This cortex then passes neural impulses to the hypothalamus and thalamus of the so-called limbic system in the interior of the brain and also to the orbitofrontal cortex in the frontal lobes of the brain. The latter location is responsible for decision-making. All of these connections more than likely evolved as a means for rapid decision-making of organisms based on olfaction.

The taste receptors in the papillae of the tongue bind to the chemicals of food or beverage we put in our mouths. Such binding then sends action potential to the brain by the cranial nerves. Although these pathways are not as direct to specific areas of the brain as for smell, they reach some of the same regions of the brain that olfactory impulses do. Three major cranial nerves take the neural impulses generated by taste receptors from the front two-thirds of the tongue, and a single cranial nerve transmits the information from the throat, the top of the mouth, and the back third of the tongue to deep into three areas of the limbic system, which includes the thalamus. From the limbic system the impulses are transferred back out to the gustatory cortex, where the source of the impulses is interpreted as sweet, sour, salty, bitter, umami, or some combination. The gustatory cortex is located in the orbitofrontal cortex, where olfaction is processed. This pathway partially explains why taste and smell are so closely coordinated. In fact, what we call taste is really a multisensory experience, combining smell, taste, and texture. Many researchers argue that this close integration of these three sensory pathways is the result of extreme natural selection on populations of organisms to make quick and precise decisions about the things they put in their mouths. Of course, as the senses are interpreted in the brain, this information then interacts intricately with the physiology of the organism with respect to the reward system of the brain. The mouths of organisms have evolved to be pretty sophisticated sensing organs.

Hearing starts with the intricate structures in the inner ear that produce action potential as a result of sound waves hitting the intricate device of the inner ear (Chapter 5). The rest of the system is every bit as complex as in vision (see fig. 13.1). A complete description would be impossible in a paragraph, so I simplify it here a bit. After impulses are created as a response to sound waves, they travel to a group of nerve cells called the organ of Corti and also by means of one of the cranial nerves that wires the inner ear to the brain stem, where a connection with a group of nerve cells called the cochlear nuclei is made. In addition, there are connections to the thalamus in the limbic system. There is one more specific connection, and that is to the primary auditory cortex of the brain into what is called the superior temporal gyrus (one of the convex rolls of the brain in the temporal lobe). The impulses travel to many parts of the brain for higher-order processing, such as understanding language and responding to language, as is the case for Broca's and Wernicke's regions of the brain.

Because the vestibular system interacts with very basic movements of the body to maintain balance, and a huge number of muscles are used to do this, the pathways for this sense are very complex, too. The cerebellum ultimately controls balance, so the axonal makeup of this pathway fords its way to this structure at the base of the brain. To accommodate this system there are what are called ascending pathways (taking information up the spinal cord to the cerebellum) and descending pathways (taking information from the brain stem back down the spinal cord). The process starts when one of the major cranial nerves carries the initial action potential from the inner ear to the base of the brain. Once there, the impulses go to various clumps or nuclei of nerve cells in the medulla and pons of the brain stem and in the cerebellum. The various clumps of neural cells are responsible for different aspects of balance. The nuclei in the pons and medulla connect to descending pathways. One pathway that lies laterally connects to the spinal cord and traverses the length of the spinal cord. Walking upright in a balanced fashion is an outcome of proper signaling along this pathway. Another lies medially, uses the spinal cord to travel to midthoracic regions of the spinal cord, and both controls how we move our head and balances how our eyes and heads move.

The five or more kinds of touch sense receptors in the skin (Chapter 8) are stimulated mechanically and produce action potentials that need to travel to the brain. These impulses ultimately traverse the brain to the sensory cortex, where the touch impulses are interpreted and acted on. There are three major highways to the brain from these sensory organs embedded in the skin based on what the information conveys. Touch and the sense of where our bodies are in three-dimensional space travel to the brain by means of neurons that run along the back (dorsal) side of our bodies. Major aspects of proprioceptive stimuli that are dealt with by our sense of balance also travel to the brain via the spinal cord. Impulses essential in sensing temperature and pain travel in yet a third pathway to the brain. Once the impulses get to the brain, they congregate in the primary somatosensory cortex, that region of the brain we discussed extensively in the context of homunculi in Chapter 3. Most of the rest of the story of the senses is about what is called multisensory integration, or crossmodal interactions. These interactions are important for fast, accurate, and sometimes lifesaving interpretation of sensory information.

That smell you noticed or that flash of light or the breeze hitting your arm are all complex perceptions that are processed in your brain. None of these—the overall perception of smell or sight or touch—is actually only a single sense at work but usually the product of senses interacting.

Consider touch in this context. The information gathered by our touch neural cells is sent by action potential from the various kinds of cells in our skin that detect the touch. This action potential is then integrated by different parts of the brain. The sensory cortex is intricately involved, as shown through Wilder Penfield's surveys of people during brain surgery. But more than just the sensory cortex processes touch. Our brains could easily stop at processing the touch information without complicating matters, but in a world of natural selection and genetic drift more complex things happen. To attain maximum resolution of the tactile stimulus that is important in an adaptive context, our brains incorporate more information about the touch for our species's survival. With respect to touch, it is well known that with it, brain activity is enhanced not only in the somatosensory cortex but in other places in the brain. The activation is in areas of the brain responsible for seeing and hearing, among others. The reason for this is that the signal our brains are trying to perceive from the initial touch might not be pure. By pure I mean that the stimulus might not have the right level of information for the brain to reach a reasonable conclusion about the tactile event. The original touch might be a hard collision of the skin with an object, and in this case the information going to the brain might be so chaotic that it overwhelms the brain and causes problems with the interpretation of that original touch. More likely, though, the original touch may be so light that other senses are needed to amplify the information that goes to the brain. Indeed, crossmodality is usually most important in situations where the original sensation is very weak, suppressed, or broken down. The brain still needs to interpret the signal in some way. A good starting point is to return to thinking about neural rivalries, and these exist for nearly all of the senses.

# 14 NEURAL DETRITUS

*Making Sense Out of a Noisy Environment*

*"The world is noisy and messy. You need to deal with the noise and uncertainty."* —Daphne Koller, computer scientist

More than fifty years ago, while some of the split-brain studies were being initiated, Dutch psychologist Willem Levelt published a long paper entitled "On Binocular Rivalry," about a phenomenon that had been known since the nineteenth century. It occurs when each of a person's two eyes are viewing strikingly different imagery or stimuli. The person observing the images can perceive only one of the images at a time, and perception of the image by the brain switches back and forth between the two images from the two eyes. Related to this phenomenon but somewhat different from it are optical illusions (box 14.1).

Rivalry also exists for some of the other senses. Several decades ago, researchers reported on a novel hearing illusion that they suggested was the result of auditory rivalry. The experiments that pinned down this auditory illusion involved right- and left-handed subjects. The choice of these subjects is valid, because handedness apparently indicates which side of the brain is the dominant side, and knowing the dominant side of an individual's brain is critical to how this auditory illusion is interpreted. A short description of handedness is appropriate here, and since

## BOX 14.1 | OPTICAL ILLUSIONS

One classic example of an optical illusion is the juxtaposition of two identical, solid white silhouettes of a man or woman nose to nose (see fig. 14.1; another similar illusion is also shown in the figure). The space between the two profiles is filled in with black to form an urn. All kinds of illusions like this one are obvious reminders that humans never perceive both images at the same time; rather, our perception of the two images shifts as the brain is forced to perceive one or the other. We simply cannot see both images at the same time. It is appropriate that the phenomenon was discovered during the invention of the stereoscope. Although binocular rivalry has been recognized for almost two centuries, we still do not know exactly how it works in the brain.

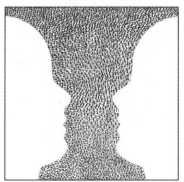 

Figure 14.1. Two classic black-and-white optical illusions. Do you see vases or faces in the left side? Do you see sexy or sax on the right?

a lot of what we know about handedness has been placed into a genetic context, I will use genetics to explain it. The genetics of some traits in humans are pushovers—they can be pinned on a single gene region that controls the expression of the trait, such as color blindness. Other traits, such as height or weight, are tougher as it turns out because multiple genes control them. Over the years many candidate genes have been proposed as controllers of handedness, but none with much certainty. Modern genome-wide association studies using whole genomes of thousands of people have been unable to pin down candidate genes, and even the tried-and-true identical twin approach has failed to distinguish candidate genes. The inability to pinpoint candidate genes suggests that many genes are involved in the expression of handedness in human populations.

Two important inferences come out of the genetic studies of handedness that are relevant to brain sidedness. First, because the trait is genetically complex and probably the result of additive effects from many genes, expression of the trait may be the result of the effects of genes accumulating to allow for a bias for right-handedness. For those who do not develop the bias, whether they become left-handed is a matter of chance. In other words, there is no so-called left-handed gene. The second point is that perhaps there is more than a single genetic way to develop this bias toward right-handedness. Nevertheless, if someone is right-handed, then most likely the left side of his or her brain is dominant, and if they are left-handed, the dominant side of the brain is the right side.

Using this trick, Diana Deutsch took right- and left-handers, placed headphones on them, and blipped sounds at different pitches into their ears. Deutsch exposed her listeners to alternating pitches of very brief (one-fourth of a second) blips of sound at 400 hertz, followed by a blip at 800 hertz, with no gap between the blips. The amplitude of the sound was equal for both pitches. The difference between the left and right ears was such that, when the left ear was hearing 400 hertz, the right was exposed to 800 hertz. These two tones were selected because the lower tone (400 Hz) is easier to hear than the higher tone (800 Hz). If perceived correctly, one should hear the high pitch in one ear and the low pitch in the other ear. And it should shift back and forth from the left ear to the right. It should sound like an alternation between "whoo-hoo, hoo-whoo, whoo-hoo, hoo-whoo," where the first "hoo" or "whoo" is in the left ear and the second is in the right ear. There should be an undulation between ears.

Surprisingly, most people simply hear the one tone in one ear and the second tone in the other ear, and the tones alternate. So, it sounds like "whoo" (right ear), "hoo" (left ear), "whoo" (right ear), "hoo" (left ear), and so on. A good number of individuals simply heard "whoo" (right ear), "whoo" (left ear), "whoo" (right ear), "whoo" (left ear), and so on. And a small percentage heard a third ghost (nonexistent) tone interspersed with the two real tones. Of the eighty-six subjects in the study, not one perceived the correct pattern of tones or their undulation. But what about handedness? It turns out that right-handers localized the higher tone to the right ear and the lower tone to the left. Even when the earphones delivering the tones were reversed, they still heard the tones this way. Left-handers did not localize the sounds to either ear and settled more or less randomly on the ear that "heard" the higher tone and the ear that "heard" the lower tone.

So, it seems that any structures on our bodies that have two entries for the sensory stimulation, or in other words are bilateral, have this rivalry. Bilateral symmetry evolved hundreds of millions of years ago, and most higher animals are great examples of such symmetry.

What other sensory organs are bilateral? Three come to mind—one obvious and the other two not so much so. The obvious one is the olfactory primary sense organ, or nostrils. One less obvious one is the touch organs placed symmetrically on the left and right sides of the body. And another less obvious one would be the balance organs.

Olfactory or nasal rivalry has recently been examined by exposing one nostril to phenylethyl alcohol (PEA) and the other to $n$-Butanol, two chemicals that smell quite different. PEA is a pleasant-smelling compound with hints of roses and honey, but $n$-Butanol has a sharp smell —a little like sniffing a magic marker. Because the olfactory system can adapt to odorants very quickly (in about twenty seconds—neither vision nor hearing have this propensity to adapt), the simple binocular and biauditory experiments discussed earlier were modified to minimize the contribution of adaptation to the results. It turns out that there is also a nasal bilateral rivalry. Specifically, when the nostrils of subjects are presented with these two different smells, only one is detected at a time, and the detection alternates much like the auditory and visual systems from one side to the other.

Processing touch is a little different, even though the tactile sense organs are in general bilaterally symmetrical. But research has revealed that the processing of touch stimulation requires not only receiving and transmitting the location of the tactile stimuli to the brain, but also that the locality be combined with information about the current posture or spatial position of the regions experiencing the tactile stimuli. The latter requirement is fulfilled visually. With crossed hands, the processing starts with the brain assuming the usual orientation of the hands—that is, left hand on the left and right hand on the right. But since the hands are really crossed, this position reverses the perception, and someone given tactile stimuli to the hands in a crossed position will immediately perceive and interpret the reverse orientation of the stimulus. As the brain realizes that the arms are indeed crossed from visual information, the tactile information is effectively remapped. The lag time between getting the orientation of the stimuli initially wrong to remapping and getting it right is a critical measure of how the remapping works. Experiments with crossed fingers indicate a similar mapping deficit, but no bilateral anatomical structures are involved in that response. One critical difference between the response of crossed hands and crossed fingers is that with crossed hands the deficit decreases with longer and longer stimulation times, while crossed fingers do not improve even with stimulation lasting almost to a second. This lack of improvement is interpreted as a basic difference between the bilateral hands and the same-side fingers.

The phenomenon has been named the "crossed hands deficit" (fig. 14.2), and it is also evident when the legs are crossed with respect to stimuli to the legs. Additional rather odd experiments have been accomplished to delve into this response. First, the mix-up in temporal order of the stimulation occurs only from the point of crossing out to the distal tip of the limb being crossed. Any location of the stimuli on the limb proximal to the crossing does not experience the phenomenon. Second, the phenomenon can be corrected by having the subject view uncrossed rubber arms instead of their own crossed flesh-and-blood arms. Potential differences exist between the sexes in how the crossed hands deficit works. Because women have been shown to be more visually dependent than males at a specific test called the rod-and-frame test, female visual dependence in spatial matters might mean that they respond differently than males to the crossed-hands deficit effect.

The rod-and-frame test has been used for more than forty years to assess the perception of verticality as a function of visual orientation. In this test, a rod is pictured within a flat, square frame. Both the frame and the rod can be rotated. When the rod and frame are positioned such that they are both perfectly vertical, the observer perceives the verticality of the bar correctly and perception of the true verticality of the rod is not a problem. An illusion is created with this system by rotating the frame away from being vertical to the observer's field of vision. What happens is that the perception of what is vertical is mixed up. The rod can be perfectly vertical without the frame, but the presence of the frame makes

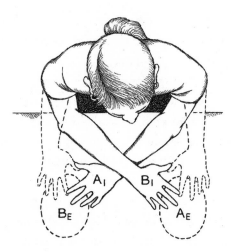

Figure 14.2. Crossed hands deficit. When the subject crosses her arms and is stimulated on one or the other hands (B1), the stimulation will appear to emanate from the wrong hand (AE) and vice versa.

the rod appear to be tilted. If the tilt of the frame is increased, the illusion exaggerates the tilt of the rod. The test asks the observer to tilt the bar so that it is vertical, and the experimenter can measure the impact of the illusion on different observations and with different tilts of the frame. Women are consistently more reliant on their vision to perceive the illusion and correct for it. Because the test requires the hands, it can be administered with crossed hands as well as with the hands in usual positions. The easiest explanation for this sex difference might reside in the differences in how males and females resolve spatial problems.

So far, we have looked at conflict or rivalry within specific senses. When we start to examine how the senses interact with one another, a whole new set of rivalries and solutions to conflicting signals arise. And even more complex are the conflicts and rivalries created by processing the senses from outside world stimuli with the higher functions of our brains like memory and emotions. A lot of important studies have been accomplished to pin down the multisensory perception mechanisms that exist in our brains. The pathways of the signals in the brain from a single sense such as touch or sight are fairly well understood. These pathways indicate a complex route from the sense organs like the eyes or ears to and through the brain that result in perception and are the best evidence for the crossmodal nature of sensing.

Precise description of these pathways involves knowing the many parts and areas of the brain. I am one of these people who tell the cabdriver, "Just get me there, I don't care how." Along the way I see landmarks I recognize, but the specific details of the trip are not so important to me. Sometimes I pay more for my ride than I would care to, but when you think of it, evolution kind of works that way, too. I will take this "just get me there" approach in attempting to describe crossmodality of senses and dispense with a lot of brain anatomy in the process. Before going into too much detail about crossmodality of the senses, I will give away some of the story. There are people who routinely can smell shapes, hear colors, and taste sounds (among other sensory mixing). These people have a rare connection of their senses called synesthesia. It will become very surprising, though, as I describe crossmodality in people not normally considered synesthetes—that is, the majority of you who are reading this book.

With five senses (visual, auditory, tactile, olfactory, and gustatory) and balance (vestibular) thrown in for good measure, there are fifteen different pairwise sensory interactions. The interactions of all of these pairs have not been fully studied and hence understood, and some are better worked out than others. If we start to examine more complex interactions such as the crossmodality of three senses, the subject gets very messy. It would be a long, repetitive chapter if we examined all possible crossmodalities. So here, to make the point of the capacity of the brain to process in crossmodal fashion, we'll stick with some of the more interesting binary ones. My favorite science center exhibits (and indeed we included one in an exhibition on the brain at the American Museum of Natural History in 2013), concern the visual-auditory crossmodality. In the AMNH experience, you walk up to the exhibit to see a life-size picture of a woman with an umbrella standing on a rainy street corner with rain pouring down on her. The visual stimulus is supplemented by nearby sound, the patter of rain falling on the street. Or is it? As you walk to the back side of the exhibit, the real source of the sound is revealed. It actually comes from bacon sizzling in a skillet!

Psychologists actually have a test that mimics this frying bacon–rainy day illusion, and they use it to tease apart the very source of visual-auditory crossmodality. One called the double-flash illusion is simple: an observer is presented with a single flash of light accompanied

A lot is known about occipital alpha waves because they are the most evident physical waveform that can be measured in the brain. This wave function is a useful property of the brain that has been used to measure many properties of neural function. Specifically, the amplitude or strength of the alpha wave has been used to measure involvement of the cortex in tasks. As we have seen in previous chapters, though, waves also have phases, and it turns out that alpha waves have a phase distribution of 100 milliseconds. Researchers hypothesize that each alpha wave that pulses through the brain delivers to the brain specific information that needs to be processed. The information so delivered cannot be augmented until the next wave comes through 100 milliseconds later.

by two auditory prompts (beeps) in rapid succession. Most people when presented with this test will perceive two flashes of light instead of one. The time between the two auditory pulses is a critical factor in determining whether the illusion will occur. When the time between flashes is very brief (less than a hundred milliseconds), the subject is more likely to observe the illusion of two flashes. The illusion stops working at a hard cutoff of a hundred milliseconds, and this suggested to researchers that there might be a universal factor in how sight and sound interact in the brain. Roberto Cecere, Geraint Rees, and Vincenzo Romei examined the possibility that visual perception in the real world is the result of an integration of information from multiple senses where information is weighted in real time to produce "a unified interpretation of an event" as a function of the brain and what are known as alpha waves (box 14.2). Cecere and colleagues suggested that the reason that two flashes of light are observed when the auditory beeps are less than a hundred milliseconds apart is that the auditory stimuli are contained within a single alpha-wave phase and the brain rushes to make a conclusion about the visual and auditory rivalry before the next alpha wave passes through the occipital lobe.

There are two possible reasons for this odd behavior of our brains. Jess Kerlin and Kimron Shapiro suggest that one reason would be an "unfortunate consequence" of the evolutionary process—that being the connection of hearing to seeing in the brain that leads to an artifact of visual perception (the second flash). The second possibility is a much more complex one with respect to thinking about how our brains work. Kerlin and Shapiro invoke a functional reason for this illusion by suggesting that our brains are performing a probabilistic analysis of the information along the lines of one proposed by the Reverend Thomas Bayes, an eighteenth-century English pastor and dabbler in probability. In his own words, Bayes recognized that the probability of something happening is "the ratio between the value at which an expectation depending on the happening of the event ought to be computed, and the value of the thing expected upon its happening."

This quaint eighteenth-century wording simply suggests that, when assessing the probability of something happening such as a flash of light, one needs to compute the impact of the knowledge of the information about an event based on observations and some prior knowledge of it happening divided by the impact of the thing happening. The contemporary language I have used might not help much, but the bottom line is that prior knowledge of something happening becomes very important when thinking about the world in a Bayesian context. So, what the brain is doing, according to Kerlin and Shapiro, is taking the Bayesian approach by using the prior knowledge that "audio beep equals light flash" to estimate the event probability that a second beep will be accompanied by a flash of light. Apparently, the prior probability of a flash of light accompanying a beep is sufficiently high to convince the brain to interpret every beep as being accompanied by a flash of light. When the beep and the flash are disconnected (more than a hundred milliseconds apart), the prior probability is perceived as low and the brain computes the probability of the event as being low based on this prior probability. In this Bayesian scenario, the brain is continuously making probabilistic statements about events occurring, and when we are left hanging for more information (for example, in the middle of a hundred-millisecond phase of the alpha wave), we use this probability to shape our perception of events. I am not sure which story I like the best. The "unfortunate" version, to me is not as unfortunate as Kerlin and Shapiro suggest. To me it is just evolution, and it shores up how much of a kluge the brain really is. But if the brain is really performing Bayesian analysis to make decisions about perception, that would be very cool indeed.

I have already discussed some of the crossmodal interactions of touch

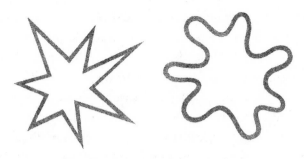

Figure 14.3. Kiki and Bouba. Guess which is which.

and sight with the crossed hands deficit, but tactile crossmodal interactions with other senses exist. Perhaps the most famous but nevertheless awesome example of tactile auditory crossmodality involves the characters Kiki and Bouba (fig. 14.3). The crossmodality of sounds with shape or texture goes further than Kiki and Bouba. When anthropologists examined the names of creatures in the languages of indigenous peoples, they discovered a surprising association of soft-sounding names using soft consonants with benign small animals and soft-leaved plants and the association of hard-sounding consonants in the names of dangerous or predatory animals and prickly plants. And indeed, as we will soon see, tastes can also be associated crossmodally with word sounds and shapes.

Kiki and Bouba are line-drawn sculptures with projections from a basic central body. Kiki differs from Bouba in that it has sharp projections emanating from it, whereas Bouba has blobby amoebalike projections. Kiki is sharp looking, and Bouba is soft and blobby. When presented with pictures or three-dimensional images of these two characters and asked to connect the name to the character, the grand majority of people associate the name Kiki with the sharp-appendaged character and the name Bouba with the amoebalike character. The association is not age dependent, nor is it culturally or language dependent. People of all ages get it, and people from different countries get it. The sharp sounds of the Ks and the soft sounds of the Bs are associated with the shapes of the characters most likely through a crossmodal process in the brain.

Not only do sounds like Kiki and Bouba influence our perception of touch, but sight also has this effect. When we see something that we want to touch and pick up or perhaps walk on, there is a definite advantage to judging its texture before we go ahead and make the attempt. A slippery piece of food is lost when we try to pick it up in a lackadaisical manner, and our necks get broken when we try to walk on a slippery surface and fall. Hence, for adaptive reasons making a tactile judgment from visual stimuli might very well be an important aspect of a primate's survival. Glossiness of objects can be used to make this visual assessment of texture and has been used as a visual tool for studying the interaction of tactile perception and visual cues for nearly a century. The Ingersoll Glarimeter was first marketed in 1922 to assess the glossiness of paper and was also used to measure glossiness of objects in psychological experiments back then. Researchers realized that glossiness was more complex than just making a single measurement, and studies using glossiness subsided until recently.

It is now known that the perception of glossiness is a complex interaction of tactile expectation and visual stimulus. Using a device that can vary the glossy appearance of objects as well as their slipperiness, Wendy Adams, Iona Kerrigan, and Erich Graf performed experiments to uncover the role of glossiness in the perception of tactile stimuli. These researchers were able to pair gloss with slipperiness across a continuum, with no gloss–no slip on one end and extreme gloss–extreme slip on the other. The results of the study indicate that participants integrate gloss level with the slipperiness level of objects. Specifically, people can detect increases in glossiness when slipperiness is increased. The converse experiment, where glossiness is decreased and slipperiness is increased (the counterintuitive situation), produced a low level of change in perception. In other words, the counterintuitive extreme when presented to a person is not perceived as different as when slippery is presented with glossy. It is as if the brain is integrating the cues from gloss and slipperiness to come to some probabilistic conclusion (hello, Reverend Bayes) that then biases the perception of slipperiness and glossiness.

The interaction of sight, sound, and touch that I have discussed are pretty obvious. But there are also interactions and crossmodalities that involve balance, smell, and taste. For balance, vision plays a huge role, but not always. Remember the spinning figure skaters who sometimes use vision to reorient the information from their vestibular system with respect to head motion (see Chapter 7)? This visual-vestibular cross-

## BOX 14.3 | GALVANIC VESTIBULAR STIMULATION AND BALANCE

Using galvanic vestibular stimulation, researchers can either amplify or dampen the response of the vestibular system during head movement when the body sways. When a person is balanced, there is a typical and measurable mechanosensory response of the vestibular organs in the inner ear. By putting helmets that can dampen or enhance the signal of the vestibular system to the brain on study subjects, researchers can induce in the subjects an illusion that while balancing their bodies their heads were moving faster or slower than they really were. The researchers could quantitate both the extent of the response by the subjects with and without vision and the intensity with which people will attempt to compensate for the illusory movement. Sounds like a really cool carnival ride to me, but the results from the study were very illuminating as to how the vestibular system sometimes is on its own even with information from visual cues. Simply put, for naive spinners, vision doesn't help.

modality, however, works only in certain circumstances. Recent experiments can actually tease apart some of the role of vision in balance and how the mechanosensory information from the vestibular system in the inner ear is cross-modulated by vision by using experiments with galvanic vestibular stimulation, or GVS (box 14.3). Normally, if a person is conditioned to swaying, he or she will reweight the information from the vestibular system and rely more on the visual system. The individual would then subconsciously recalculate the position of the body axis that maintains balance based on the downweighted information from the vestibular system and the upweighted information from the visual system. It's kind of like figure skaters practicing spins over and over to condition themselves to the amazing spinning moves. If they aren't conditioned and naively spin, this action wreaks havoc on the skaters' sense of balance until they have practiced the moves repeatedly. The naive spinners' brains do not have the capacity to receive the reweighted vestibular and visual information, and hence the vestibular system appears to be entirely on its own.

We are bombarded by sensory input all the time, much of it information that is superfluous or junky and what I call neural detritus. If we processed every bit of sensory information to which we were exposed, the brain would be an overworked mess. Shortcuts abound in how we process sensory information, and this is because as the action potential from the external stimulus comes to our brains we have evolved specific ways to sift through the noise and interpret the information coming in. For example, most humans are pretty adept at focusing on a conversation with someone even in a crowded, loud room. On the other hand, often we need to make quick decisions based either on paltry sensory information or on loads of conflicting information. For instance, many optical illusions are the product of the brain simply saying, "I give up, but here is the best I can do." This chapter focused on several phenomena used to explore this interesting aspect of our ability to sense the outer world and the brain's capacity to filter out the sensory detritus. This capacity is what makes us able to perceive the outer world as orderly and concisely as possible.

# 15  PANI CA' MEUSA, CRÈME BRÛLÉE, AND SYNESTHESIA

*Crossmodal Impact on Taste and Synesthesia*

*"Crème Brûlée is the ultimate 'guy' dessert. Make it and he'll follow you anywhere."* —Ina Garten, cookbook author and host of *The Barefoot Contessa*

Taste is based on chemoreception, but it turns out that what we taste is heavily influenced by our other senses, such as sight and touch. On a visit to Palermo, Sicily, I had the pleasure of eating street food at the famous Ballarò open-air market. I stumbled on a vendor preparing and selling a Sicilian delicacy called pani ca' meusa. Believe me, it did not look appetizing at all. That it seemed to be simmering in a laundry bucket made it look even more unpleasant. The color of the "delicacy" was a sickly gray with a greenish hue, and the texture of the meat in the sauce was, well, porous and spongy in places and rather jagged in others. I am being delicate here when I say it didn't look edible, and when pani ca' meusa was translated for me (bread with spleen), I was even more turned off. Then I learned that lung and throat cartilage were also elements of the sickly gray sauce, and I could make out the rough cartilage and even spongier lung particles in it. At this point I was massively second-guessing having ordered this so-called delicacy. My first taste of this street fast food, not surprisingly, gagged me a bit, because it was sour and had a lumpy texture. But I conducted a simple experiment. I closed my eyes, tried to put my preconceived notion of what a spongy gray food might taste like out of my mind, and took another bite. Not much better. I finished the pani ca' meusa reluctantly. Later that week, though, after my preconceived notions had subsided, I tried it again. I can now say that pani ca' meusa is perhaps one of the neatest, tastiest fast foods I have ever eaten and crave returning to Palermo's street markets for more whenever I think of Sicily.

Part of the effect on perception of taste also depends on the semantics of description of the food. Actually, pani ca' meusa has a pleasant-enough sound in Sicilian, so here is where I will momentarily drop the bread and spleen. Marketing researchers have started to use the intricacies of crossmodality to enhance sales of their products. Using experiments that suggest that a product's name is incredibly important in how the product is perceived, marketing researchers have discovered some universals. One of the most amusing is that harder-to-pronounce wineries are perceived as making better-tasting wine than less-difficult-to-pronounce wines in head-to-head competition. So, the language used to describe food can influence taste greatly. Connect this factor with the tactile crossmodal interaction with taste, and we have a very complex way to perceive the taste of our food. We have seen that tactile information or texture of food is important to perception of taste. This also might mean that shapes are important in perception of food, and indeed research shows this very response.

Psychologist Charles Spence has made a career of understanding the crossmodality of taste with other senses and is described as a pioneer of taste studies. He has studied the speech sounds associated with chocolate, a great example of his interest in crossmodality and taste. Spence and his colleagues looked for any associations between chocolate and its cocoa content with rounded and sharp or angular words in different cultures, kind of like a taste-based Kiki and Bouba. They found that sweet or low-cocoa chocolates are almost always associated with round words with sound containing soft nasal and back vowels. Words such as "lula" and "maluma" are appropriate examples of these soft- or round-sounding words that are associated with sweetness. Bitter chocolate or high-cocoa-content chocolates are associated with words such as "tuki" and "takete." Although the experiments they used clearly show that people can map sounds to tastes, the exact mode of the mapping is still unknown. But given Spence's originality in experimenting and his interest in everything from the crunch of a Pringles potato chip to the fizzing sound of carbonated beverages, it would be a good bet that the crossmodal nature of a lot of these associations will be worked out soon.

Most of my original aversion to bread and spleen was visual. Specifically, warnings early in life from my parents never to eat grayish green meat probably contributed to my initial aversion to the delicacy. But another sense was probably at work as part of crossmodal interactions that caused my gag response. Researchers think there is also a tactile component to what we taste. Experiments where subjects are given food objects that are associated with specific tactile influence can address this possibility. In experiments, subjects were served the same food in two ways: with a rough surface and with a smooth surface. They also received the same food on rough or smooth platters. The researchers used the serving plate experiment as a control and indeed found that the texture of the serving plate had no effect on gustatory reception. The texture of the surface of the food, however, had a significant impact on what study subjects were tasting. Whenever a person tasted the food with a rough surface, he or she perceived it as sourer than the same food served as smooth. Perhaps the rough texture of the throat cartilage mixed into pani ca' meusa induced a similar sour taste.

The next question is: Does language based on shape act in a crossmodal way with taste? The answer is yes. In clever experiments, researchers determined that the semantic space of taste and shapes contains two principal clusters. This is just a fancy statistical way of saying that their experiments revealed two associations of words and shapes. One cluster consisted of the word "sweet" and round shapes. The other cluster included shapes that had edges and the taste words "salty," "sour," and "bitter." Using these associations, the researchers timed study participants' response to incongruent and congruent pairings. This test is kind of like a shape/taste Stroop experiment (box 15.1). In this Stroop taste test, a series of food items, each having a specific shape and a specific taste, is presented to the person being tested. So, for instance, the person might be given a series like the following: a round/sweet piece of food, then a jagged/sour piece, and finally a triangular/bitter piece. The subject is asked to name the tastes in the series and is timed doing the task. More

The Stroop test is a classic vision-semantic test that pairs words with color in a unique way. It consists of two lists of words in which the words are printed in different colors. In one list the letters are colored the same color as the word, so BLUE would be blue in color, GREEN would be green in color, and so on. The second list mixes up the colors and the words, so that BLUE would be colored red, GREEN would be colored purple, and so on. The person taking the test is asked to verbally list the colors of the words. One can easily and quickly list the colors of the words when the name of the color is matched with its visual color. But to call out the colors of the words on the list correctly when the colors don't match the word takes more effort and much longer. The Stroop test is a classic example of a "team of rivals" effect. When we see the word RED colored blue, this produces a conflict in our brains that requires sorting out, which takes some time. When we see the word RED colored red, there is no conflict, and hence we can quickly make the decision that the color of the word is red.

shape and taste combinations follow, such as star shape/sweet, followed by oval/sour, followed by round/bitter. The subject is again asked to identify the tastes. When the shape and taste combinations are incongruent —for example, star shape/sweet—it takes longer to identify the taste correctly. This result indicates that there is a crossmodal connection of taste and shape as mediated through language.

Words are obviously part of these crossmodal interactions. But what about other sounds we make, such as music? One need look no further than the music in TV commercials to conclude that music might be importantly linked to our perception of taste. It could be argued that there is some aesthetic component to the choice of music in a TV commercial, if one can imply that a commercial has aesthetics. But there might also be some very distinct crossmodal interactions going on in such choice. Now we can go one step further and determine whether music is involved. Enter Charles Spence once again. With several colleagues, he has examined the potential role of music in perception of taste. In a series of experiments, Spence and his colleagues showed that sweet and sour tastes are associated with high pitches, whereas bitter tastes are linked to low pitches. Even musical instruments elicit connections to tastes. For instance, piano sounds are linked to sweetness, and trombones map to bitter and sour tastes.

These results simply ask what taste words are associated with basic units of sound or music. Digging deeper into the crossmodal role of music, Spence and colleagues looked at musical composition as a potential player in crossmodality. They asked a sound-branding agency to create four musical pieces that varied in auditory pitch, sharpness, roughness, and discontinuity. Using what is called a forced choice experiment, these scientists then asked participants to assign each of four taste words to the four compositions. The forced choice aspect of the experiment was that they were instructed to match one of the taste words with one of the compositions so that there were four unique matches. The researchers also assumed from their first experiment that there were correct assignments, such that sweet would be paired with a higher-pitched, softer, more continuous, and less sharp-sounding composition. And the other taste words would be associated with appropriately manipulated compositions. The results indicate that sweetness can be matched to its "correct" composition significantly better than randomly, with the salty taste word to its correct composition slightly less accurately, and the bitter and sour taste words least accurately. Both of these experiments support a hedonic model of crossmodality. In other words, the pairing of sweetness with more continuous and less sharp music can be based on the pleasure factor of the taste and the sound. In general, moderate sweetness is more pleasurable to the brain than bitterness, sourness, or saltiness. Likewise, more continuous and less sharp music is more pleasurable to people than discontinuous, sharp-sounding music. The overlap in pleasure level from each might be driving the capacity to pair sweetness with the more pleasurableness of the music.

Another experiment Spence and colleagues designed was relatively complex, because it paired four tastes with six kinds of music, for twenty-four possible combinations of music with taste. The six kinds of music involved subtle changes using attack, discontinuity, pitch, roughness, sharpness, and speed. For instance, pitch could be varied by the subject on a scale they controlled from high pitch to low pitch. The subjects were shown a single specific taste word ("sweet," "sour," "bitter," or "salty") and then asked to find where on the sliding scale for the six aspects of music the word could be matched. So, for instance, for the word "salty" the subject would read the word and would find the pitch that matched the word. Next, they would see the word "salty" and then find where on a sliding scale speed matched the word, and so on, for all four taste words and all six aspects of music. Surprisingly, all but attack of the music showed significant mapping with taste words. And in some cases, musical aspect can rank the taste words. For instance, auditory or musical roughness was significantly correlated to taste words with "sour" being associated with very rough sounds, "salty" with medium rough sounds, and "sweet" with the softest sounds. In general, sweet is mapped with higher pitch, softer sound, more continuous sound, lower tempo, and less auditorily sharp sound. Sour, on the other hand, can be paired with low pitch, soft sound, discontinuous sound, high tempo, and sharp sound.

The final experiment Spence and his colleagues designed considered the possible cultural differences that might be involved in the crossmodality. They recruited subjects from the United States and India and repeated the forced choice experiment with the four musical compositions and four taste words. The decision to contrast Indian subjects with American subjects was based on the cultural context of music preferred in the two countries. Indian music is microtonal; it uses intervals smaller than a single semitone. Western music as preferred by people from the United States uses a specific twelve-semitone interval system. This may not sound like much when read on the page, but the differences in music caused by these two profound structural preferences ensures quite different musical perception, preference, and processing. The results of the experiments indicate that American subjects are better at matching the "correct" composition with the taste word, suggesting that a cultural component is involved in the crossmodality. Besides telling us something about crossmodality, these experiments have an obvious marketing function. They point to what the Internet calls the now popular "pretty stupid yet insanely cool" Häagen-Dazs app. This app plays a classical concerto as a timer as you wait for your ice cream to breathe before eating it. It appears that Häagen-Dazs could have chosen the concertos for their ice creams a little more judiciously if they had followed Spence and colleagues' advice on how to pair music with sweet taste. The concertos on the app are apparently not optimized for pairing sweetness with composition characteristics. On the other hand, rap music might be a good way to advertise salty products because of the pairing of discontinuous choppy music with salty taste words. And most important, if you are an Indian marketer, you wouldn't want to pair a classical concerto with the sugary dessert rasgulla in TV advertisements.

Smell can also be crossmodal with sound and even music. Again Charles Spence, the guru of this kind of work, and his colleagues have studied the pairing of smell with sound. To test the validity of the olfactory-music crossmodality, these researchers used previously demonstrated olfactory-included crossmodalities. For instance, it is well known that smells are paired with shapes. Spence and colleagues were able to pair such smells as crème brûlée more with rounded shapes than with musky odors. Using the strengths of these crossmodalities, they then assessed the strength of any olfactory-music crossmodalities. In their first set of experiments, sweet, round, and higher-pitched correspondences were detected. As with the taste and music results, the response is more than likely hedonic, since sweet tastes, roundness, and higher pitch are all associated with pleasantness. A second set of experiments paired musical compositions with three smells: candied orange, crème brûlée, and ginger cookies. Most people in this study easily paired the candied orange smell with its "correct" music. On the other hand, people would randomly pair crème brûlée with all three musical compositions. And very strangely, ginger cookies were never matched to the "correct" composition. These results are revealing of a crossmodality of olfaction and sound, and specifically with music, but as the response to ginger cookies indicates, the interactions are most likely very complex.

Thus far we have examined ten of the fifteen possible binary crossmodal interactions for six senses (sound and sight, touch and sight, touch and sound, balance and vision, taste and sight, taste and sound, taste and touch, smell and touch, smell and sight, and smell and sound). We are missing four of the balance-with-other-sense pairs and one other pair: smell and taste. The four balance pairs that are missing would indeed be difficult to show existed, but I would not be surprised if Charles Spence figures out a way to pin these down if he puts his mind to it. The remaining pair is smell and taste. These two senses are well known to be intricately bound together, and their crossmodality is not in doubt. We have also looked at some three-sense crossmodalities, such as taste, sound, and sight. But the message should be that a team of rivals is

Figure 15.1. Nigel Tufnel's famous amplifier dial that "goes to eleven" and is "one louder."

taking the sensory information generated by our sense organs and interpreting it.

Before we leave crossmodality, I want to return to Nigel Tufnel, the lead guitarist for the mythical heavy metal band Spinal Tap, mentioned in Chapter 7. His famous guitar amplifiers all had volume dials that went to eleven (fig. 15.1). When asked why not just number the dial levels from one to ten and have ten as the loudest, he responded, "Well, it's one louder, isn't it? It's not ten. You see, most . . . most blokes, you know, will be playing at ten. You're on ten on your guitar . . . where can you go from there? Where? Nowhere. Exactly. What we do is if we need that extra . . . push over the cliff . . . you know what we do? Eleven. Exactly. One louder." There is actually some crossmodal sense to Tufnel's bantering, as demonstrated by the following experiment. Subjects were exposed to a reference sound and asked to compare it to a second sound of either louder or softer magnitude. A small number (one, two, or three) or a large number (seven, eight, or nine) accompanied the second sound, and the subject was asked whether the reference sound was louder or softer than the second sound. The trick of the experiment is to expose the number to the subject simultaneously with the second sound and after a pause between the sound and the exposure to the number. When the number and sound occur simultaneously and sounds are paired with large digits, the subjects perceived the sound as louder than when sounds are paired with small digits. By separating the sound from the number though, the effect disappears. In this case seven, eight, and nine are indeed louder than numbers smaller than them, and if that is the case, then eleven could be louder than ten, too.

I have glossed over one important point about all of these experiments on crossmodality. Synesthetes were excluded from all of them. These phenomena are shown to exist in individuals without obvious connections of senses, as synesthetes clearly show. And synesthesia has added a great deal to our understanding of crossmodality and, more important, to the route and mode by which action potentials originating from our sense organs (eyes, ears, or noses) traverse our brains. Synesthesia, like the crossmodal interactions I have just discussed, comes in many varieties. According to many synesthesia experts, there are probably between 65 and 150 kinds of synesthesia. Perhaps the most accurate estimate is 80, as listed by Sean Day. To cover all 80 would be a book in itself. So I will touch on several subjects where some of the 80 types can help us understand something about the senses and, more important, help us uncover something about how our brains work to create perception and, ultimately, consciousness.

Each kind of synesthesia is made up of an inducer and a concurrent. The inducer is comparable to the trigger for the synesthetic experience. Without it, the synesthetic response cannot occur. If it is present, then the inducer does what its name suggests and induces a sensory response that is not a part of the common response specific to the inducer. The concurrent is the sense or sensory response that results from the presence of the inducer. For instance, the most common synesthetic response is where numbers or letters (also known as graphemes) act as inducers and the concurrent is color. Over 60 percent of people with clear synesthetic abilities have this grapheme-color pairing. Indeed, the clearest way to determine if someone is synesthetic is to test for this pair. It also turns out that if someone is synesthetic for a particular inducer-concurrent pair, he or she might very well be synesthetic for another or even several other pairs.

Synesthesia was at first thought to be more prevalent in females. The ratio of female to male synesthetes differs depending on which expert is consulted but ranges from six to one all the way down to one to one, where there is no female bias in the trait. Sean Day suggests that synesthetes make up about 4 percent of the human population. But this number will vary, too, depending on the study and researcher. Part of the difficulty in pinning down these frequencies of synesthesia in human populations is that you simply can't use self-reported synesthesia in making the estimates. Consequently, several clever tests of whether a synesthetic claim is genuine have been developed. A modified Stroop test is one of the basic tools used to detect synesthesia. The only problem with the Stroop test is that some nonsynesthetes can train themselves to overcome the propensity to use the word they are seeing to describe the color and therefore appear to be synesthetic. Some individuals have also trained themselves either purposefully or simply by life exposure to inducer-concurrent pairings and hence can "cheat" their way to being considered synesthetes. The most visible of such false synesthetes are people who remember the colors of refrigerator magnet letters from childhood and retain the association of the magnet color with the alphabet letters. Children's alphabet books can elicit the same false grapheme-color synesthetic response.

The most widely used test for synesthesia is the test of genuineness (TOG) and its improvement, the revised test of genuineness (TOG-R). These tests rely on the ability of the potential synesthete to systematically repeat the synesthetic response when challenged at different times. So, for instance, the potential grapheme-color synesthete is presented with a long list of graphemes that include words, days of the week, numbers, and letters of the alphabet. After the mention of each grapheme, the subject is asked to describe their precept of the grapheme. Different TOGs will use different inducers, but the real meat of the test lies in the repeating of the graphemes at different points. The TOG-R is a more complex and more precise method of testing for synesthesia that has also been cross-validated by applying the test to a large sample of known synesthetes. The Stroop test, mentioned earlier, has been suggested as a potential way to detect synesthesia. Remember that this test causes confusion when nonsynesthetes attempt to assess color in conflicting situations. When the nonsynesthete is challenged by words like "blue" printed in colors other than blue, such as green, they will confuse the color of the word and use the word itself, in this case "blue," to describe something that is green. Synesthetes will in general not be fooled by the Stroop test.

Synesthesia has been studied by scientists for about two hundred years. The first study of synesthesia, published in the 1820s, happens to have been conducted by a synesthete whose dissertation was in part a study of himself. The phenomenon piqued the interest of researchers but didn't really get the bump it needed until Francis Galton, Charles Darwin's cousin, first looked at the psychometrics and possible genetics of synesthesia in the late 1800s.

Galton was a famous figure in Victorian scientific circles. He was a polymath who was difficult to categorize by his specialty and a brilliant statistician, given the mathematical tools of the time. Psychologists often times claim him as one of their own because he invented many of the early statistical tools that were used in the science of psychometry popular at the time. Synesthetes were of particular interest to Galton as a psychologist, and he was the first to show that synesthesia ran in families. He also showed clearly that not all synesthetes see the same color associations when challenged with identical graphemes. He also championed genetics as a scientific discipline. Unfortunately, he also defined and championed the now-disavowed pseudoscience called eugenics. Although Galton gave eugenics its name, he didn't name synesthesia, even though the phenomenon was a subject of great interest to him. Rather, that distinction belongs to Jules Millet, who named the phenomenon in 1892. Erica Fretwell, an expert on late nineteenth-century literature and culture, has pointed out that synesthesia and eugenics were bound together, partly because of Galton, but also as a result of contemporary social mores. Victorian intellectuals were hung up on the differences within our species. Synesthesia offered yet another set of criteria for characterizing differences, and because it was neural or brain based, it linked more than genetics to these differences.

Binding synesthesia to eugenics was an unfortunate association for synesthesia. But since the demise of eugenics, the study of synesthesia has opened up many avenues of research on neural processes, and there should be no guilt by association of synesthesia with eugenics. Nevertheless, studies of synesthesia waned in the early twentieth century when eugenics went out of vogue. Synesthesia studies luckily reemerged around the late 1980s. The focus on synesthesia in a genetic context has led to several advances in how we view the phenomenon. Galton noted the familial correspondence of synesthesia but lacked the tools to pin down its genetic basis. Also, the sex bias that has often times been associated with synesthesia has misled researchers for some time regarding the location of genes involved in synesthesia. If the sex bias is as high as some studies suggest, then this might mean the trait is linked to the X chromosome, much like the X-linked tetrachromatic vision characteristics discussed in Chapter 3. The early observation Galton made about synesthesia running in families has also come into play in trying to uncover the genetic basis of synesthesia. A form of synesthesia called color-sequence synesthesia was examined in a twin study in 2015 by researchers Hannah Bosley and David Eagleman.

For this form of synesthesia monozygotic (identical) twins have a heritability of the trait that is about 74 percent, and dizygotic (fraternal) twins have a heritability of 36 percent. If the trait were controlled entirely by genetics, then the monozygotic twins would have 100 percent heritability. If the trait were not inherited, then it should appear in either kind of twin at a much lower concordance. The 74 percent heritability of the dizygotic twins in this study suggests that this kind of synesthesia has a genetic component, but it is not total, and that expression of the trait has a considerable environmental component. For another type of synesthesia, the genome-wide association study approach is useful.

Modern genomic studies using the genome-wide association study approach exploit whole genome sequences of reference populations and the sequences of individuals with a trait of interest. These genomes are mined for positions that vary among the controls and the people with the trait, called a single-nucleotide polymorphism (SNP; see Chapter 8). When the human germline cell chromosomes replicate to make sperm and eggs, recombination, or genetic exchange of information, between the chromosomes in cells inherited from the parents occurs. These replication events break up the sequential arrangements of single-nucleotide polymorphisms with traits depending on how close the trait is to that polymorphism. If the polymorphism and the trait are very close to each other on a chromosome, then the recombination or shuffling of the trait and the polymorphism will be infrequent and the trait is said to be linked to it. Pinning down an association therefore requires showing that a specific polymorphism or multiple polymorphisms are linked to the trait. Furthermore, if we can determine where the polymorphism is on a chromosome, it is possible to associate particular genes in the region of the polymorphism with the trait.

Genome-wide association studies suggest that one kind of synesthesia (straight grapheme-color synesthesia) can be mapped to four of the twenty-three human chromosomes (none of these are the sex chromosomes, X and Y) and another kind (colored sequence synesthesia) can be mapped to chromosome 16. Of the four chromosomes in the first case that show linkage to synesthesia, chromosome 2 is the most significant. The other three chromosomes (5, 6, and 12) show suggestive linkage. To show the potentials and pitfalls of the approach, consider some of the genes near the linked single-nucleotide polymorphisms on chromosome 2 that associate with straight grapheme-color synesthesia. Looking for genes associated with a trait is a little like looking for a needle in a haystack, as we discussed with James Lupski and CMT syndrome in Chapter 8. If an association gets made, then the haystack gets reduced a lot, but some luck is still involved. Fortunately, 80 percent of the twenty thousand or so genes in the human genome have known, precisely described functions. The location of where a gene is expressed, how the protein translated from the gene works in development and in regular physiology, what the protein does in pathways, and other parameters are well known for these characterized genes. So, what kind of genes should we be looking for near the single-nucleotide polymorphisms that are associated to the trait? One obvious category would be any gene involved in how our nervous system works or genes that affect the development of our nervous system. Another category might be genes linked to neurological disorders or other neurological anomalies.

It turns out that the polymorphisms linked to synesthesia on chromosome 2 are in the same region as genes linked to autism. People with autism experience sensory-related abnormalities, and synesthesia is often a secondary feature of autism spectrum individuals. Indeed, people with autism spectrum disorder appear to have an increased frequency of synesthesia. Finally, using functional magnetic resonance imaging (fMRI), researchers have shown that auditory stimuli will excite similar auditory and visual regions of the brain for both people with autism and people who are synesthetes. As far as candidate genes go, a gene linked to synesthesia on chromosome 2 called TBR1 is involved in telling other nervous system genes when to express themselves. In other words, TBR1 regulates several genes important in neural development, including a gene named *reelin* (a gene involved in cerebral cortex development). Another gene involved in neural processes that is also found in the region of the genome with the associated polymorphisms is called SCN1A. This gene encodes a protein that lies in the membrane of synapses and is involved in processing action potentials across synapses. People who have altered forms of this gene suffer from epileptic seizures. Circling back to autism, researchers also know that rare variants in TBR1 and SCN1A are found in people with autism. Chromosome 16, mentioned earlier, also has single-nucleotide polymorphisms associated with colored sequence synesthesia. This kind of synesthesia, which is triggered by sequences of graphemes such as ABCD to produce colors, is quite different from grapheme-color synesthesia, so it is not surprising that it might be found on different chromosomes than the grapheme-color type. Six genes in this region are involved in the development and maintenance of the nervous system in the cerebral cortex. But when these genes were examined closely for variation between synesthetes and nonsynesthetes, no variation could be correlated to the trait.

As I was writing this book, I took my two-year-old son to a hearing specialist for hearing tests. He had just had tubes put into his ears to facilitate drainage of fluids that impaired his hearing. I was a little skeptical about the whole process of testing hearing in a two-year-old. How would they get him to indicate which ear a sound is coming from? One side effect of his hearing impairment is delayed speech, and he certainly couldn't communicate verbally the way I did during my last hearing test. Because he is a typical two-year-old, I was asked to hold him during the test. I was simply amazed by how precisely the testers could interpret his head and eye movements to indicate where he heard the test sounds coming from. Researchers using similar approaches can now test children as young as two months for various sensory perceptions. You'd think that these tests would be rather simple, but it takes clever thinking to get into the mind of a two-month-old. Of course, infants don't know what letters or numbers are, but researchers can actually do tests with the two-month-old-baby version of grapheme-color synesthesia. The test is based on the idea that a two-month-old baby can associate shapes with colors. So, for instance, if the baby associates a triangle with a red color, the baby will not clearly distinguish a triangle set on a red background but will see the triangle clearly on a green background.

The trick of the test, as with the hearing test on my two-year-old son, is to read the reaction of the two-month-old who is exposed to these figures. It turns out that a baby will stare at something interesting, and viewing figures like triangles and squares is more interesting to a baby than looking at a continuous red background. So, if the baby associates a triangle with red and a triangle is presented with a red background on the left and a green background on the right, the figure on the right, where the triangle is visible, will be more interesting and he or she will stare at it exclusively. Switching sides of the two figures should result in the baby staring to the left. Almost all babies stare nonrandomly at these figures, and they do so across trials on the same day as well as trials on different days. Two interesting secondary results from such experiments add to how a baby's responses inform us about synesthesia. First, babies who pick up the red-green contrasts grasp them much earlier in life and the yellow and blue associations a little later in development. And second, as the baby grows, the synesthetic effect tends to diminish.

In the aggregate, these data reveal that most, if not all, babies are synesthetic just after birth. This observation is called the neonatal synesthesia hypothesis. The reason babies observe red-green associations before yellow-blue ones is that in the brain the red-green connections develop first and the yellow-blue channels emerge later. The final observation of diminishing synesthesia in growing toddlers, requires that we understand something about how infant and young brains develop.

Brain development has two main phases. Early in development we are born with almost all of the neural cells we will need in life. Studies that examine the neural connections of these cells indicate they are almost indiscriminately connected to each other by producing connections in an incredibly large number through synapses. Such synaptic connections cross sensory regions of the brain and in essence connect the different senses. The second phase occurs as the baby develops and is characterized by a process known as pruning. The neural connections that implement the universal neonatal synesthesia are pruned away, leaving only neural connections that process the visual information exclusively. Other neuroanatomical and neurophysiological data support this way of looking at the neonatal synesthesia hypothesis.

Another synesthetic system, colored hearing (where specific pitches of sound are paired with specific colors or shapes), can also be examined in infants. As we saw with our discussion of crossmodality, higher pitches are usually associated with Kiki-like sharp, pointy shapes and lower pitches are paired with Bouba-like rounded shapes. With this kind of synesthesia, high pitches are also paired with smaller, "higher in space" shapes and lower pitches with larger, "lower in space" shapes. In this case, detecting the baby's response is critical, and it turns out that the test can be done because babies like to look at things. So, if a baby is shown a visual in which a shape morphs from being Bouba-like (amoeboid) to Kiki-like (sharp and prickly) and the sound accompanying the visual goes from a low pitch to a high pitch, the baby's response should be different than when the visual is the same but the sounds go from high pitched to lower pitch. Specifically, because the Bouba-to-Kiki visual spectrum paired with the low-pitch-to-high-pitch sound spectrum is congruent and the same Bouba-to-Kiki visual paired with high pitch to low pitch is incongruent, the baby will be more attentive and watch the visual longer for the first combination than the second. Another test of this kind of synesthesia in infants is accomplished by showing a baby a brightly colored ball moving up and down in space. Paired with the ball's movement are sounds going either from low to high tones or high to low tones. Again, if auditory synesthesia is at work, the baby should prefer the pairing of a ball in the down (up) position with a low (high) pitch to the nonconcordant opposite pairings. All babies as young as three to four months indeed pay more attention to the congruent visual-audio pairing than to the incongruent one, suggesting that infants are synesthetic with respect to pitch-shape and pitch-height pairings.

As I mentioned earlier, there are more than eighty kinds of cross-modal synesthesia. Going into each one would be a repetitive task, so in the rest of this chapter we will look at specific aspects of synesthesia that help us better understand the senses.

First, let's consider studies that connect synesthesia with brain anatomy. Brain imaging studies can do a couple of things. For one, they can decipher the location in the brain where specific sensory information is processed. And more important, they can lead to an explanation for why synesthesia occurs in the first place. As inferred from the infant synesthesia studies, proximity of neural connections is a factor in promoting synesthesia in an individual. In infants, the connections are made but later pruned away. In true synesthetes, the connections are not pruned away and remain into later life. The earliest brain imaging studies of synesthetes were accomplished in the 1980s by infusing radioactive xenon with oxygen, which was then inhaled by a subject who was being evaluated for synesthesia. Xenon is an inert gas and is not harmful to a person's physiology, but it can be traced using the same methods used to detect X-rays, so the subject wore a cap with an array of X-ray detector devices to measure where xenon (and hence oxygen) went in the brain during synesthetic activity. During these tests the subject breathed in oxygen and xenon, put on the detector cap, and then underwent exposure to a synesthetic inducer. Although the method could detect increased brain activity, it did so only rather crudely. These experiments (using state-of-the-art techniques and knowledge from more than thirty years ago) demonstrated that the cortex experiences increased activity, but learning how was not more precisely feasible.

With the invention of positron emission tomography (a functional imaging technique used in nuclear medicine to reveal metabolic processes in the body), MRI, and later DTI (which provides information about the brain's white matter, or axons), researchers have been able to measure brain activity during synesthetic experiences more precisely. Grapheme-color synesthesia, because it is one of the more common forms of true synesthesia, has been the focus of detailed hypothesis testing about how the phenomenon works. Researchers can accomplish this most efficiently by using imaging, specifically fMRI, which highlights regions activated by a specific function. In essence, the brain is painted in different hues according to the level of activity caused by action or functioning as a result of a stimulus. The first fMRI studies of synesthetes claimed a clear correlation of grapheme-color synesthetic activity with regions of the optic pathway that correspond with the function of processing color. Later fMRI studies were designed to test hypotheses about the nature of synesthetic activity, as contrasted with the simple brain-painting experiments that were initially accomplished.

One of the more popular and logical hypotheses that can be tested concerns the process of cross-activation, which predicts a different pattern of brain activity than other models of synesthesia. Cross-activation would produce brain activity in two different parts of the brain, one where the inducer signal is being processed and the second in the region where the concurrent gets activated. The most recently developed approach to be used to examine synesthesia is DTI, and as of 2014, fewer than seven studies had been conducted using this approach. As discussed in Chapter 4, this approach can pinpoint neural connections that are active during specific neural activity. Early results using both fMRI and DTI point to a possible importance of localized brain function and connectivity, but unfortunately, there is so much variation in synesthetic activity even among grapheme-color synesthetes that interpretation of the data is difficult.

Jean-Michel Hupé and Michel Dojat performed a meta-analysis of all of the published synesthesia brain imaging studies in the literature in 2015. Their conclusions are nicely summed up in the following statement from their work: "We did not find any clear-cut empirical evidence so far about the neural correlates of the subjective experience of synesthesia. We did not find any structural or functional anomaly in the brain of synesthetes that could explain synesthesia. In our view, most published studies to date show, in fact, that the brains of synesthetes are functionally and structurally similar to the brains of nonsynesthetes." In essence, synesthesia is very complex, and perhaps we need a new theoretical way of thinking about how the various forms of it work in conjunction with how the brain is structured and functions. Like an interesting phenomenon in nature, we always have more to learn, and synesthesia appears to be an enigma waiting to be cracked.

We very likely have more than the five Aristotelian senses plus balance on which I have focused thus far. But is it possible that the big six themselves are actually more than just six? There are multiple brain regions where the pathways for each sense traverse. Vision, for example, processes all kinds of aspects of that sense, including color, shade, orientation, movement, and others. In many ways, the most common synesthetic phenomenon, grapheme-color synesthesia, is really not characteristic of almost all other synesthesias. It is what is called an intramodal synesthesia because you need to see to visualize letters and numbers, and this requirement then induces another visual concurrence. Hearing is another, and recent synesthetic research suggests that specific aspects of hearing can be atomized. A rather famous hearing illusion, called the Doppler illusion, discovered in the 1990s, illustrates how the sense of hearing is rather tightly wired with different components—one for loudness and one for pitch.

The Doppler illusion occurs when one hears a sound that is continuously increasing in intensity over time. It can occur while you are standing still and the sound is increased at a constant rate, or it can happen if you are standing still and the sound is constant but approaching you at a constant rate. What happens when one is exposed to this kind of sound is that the tone's pitch appears to increase as the loudness of the sound increases. In this case, the pitch actually isn't increasing, but the loudness is. These two aspects of auditory sensing—loudness and pitch—compete for the right to interpret the sound. The solution is to let the loudness dictate the concurrent response in pitch. This intramodal synesthesia lies in strong contrast to all of the other eighty intermodal synesthesias and suggests that hearing at least might really be two senses.

Two final synesthetic oddities concern ideas that may eventually transcend how we normally look at the phenomenon. Both are somewhat controversial and are based on much smaller sample sizes than the better-known forms of synesthesia such as grapheme-color synesthesia. The first oddity concerns a neural phenomenon that is really in a category by itself. Twenty years ago, Giacomo Rizzolatti and his colleagues noticed something interesting about the macaques they were studying. They were evaluating neural signals in the brains of these primates as they reached for food or other objects, and they recorded the electrophysiology of single neural cells during these actions. Oddly enough, when the macaques observed a human or another monkey reaching for an object, the very same neural cells showed activity. The neurons involved in this strange phenomenon were given the name mirror neurons. Other researchers later established that there are mirror neurons not only for movement of the hands but also for facial and mouth gestures. Mirror neurons have since been used in the neurobiological community to better understand many cognitive mechanisms, such as how organisms understand goals and how organisms behave empathetically. They have also been used in studies of autism. But the most recent work on the mirror neuron phenomenon has been linked to certain kinds of synesthesia, too. In mirror touch synesthesia, individuals who see someone else being touched actually feel as if they are being touched. So this involves a visual induction, followed by a concurrent tactile sensation.

Finally, the literature contains several reports of emotional responses associated with sensations. The evidence for this synesthetic oddity is sparse, however. In one subject, certain textures elicited very specific emotional responses, such as denim being associated with depression. In the common grapheme-color association, some people showing this association will also associate emotions with the grapheme-color response. In 2013, researchers treated an individual who suffered a stroke that caused a lesion in the thalamus. This lesion resulted in the first reported case of acquired synesthesia involving emotion and grapheme-color synesthesia, because the person affected reported feeling disgusted when he read words printed in blue and less so when reading words in yellow. Even more interesting was this subject's response to brass musical instruments (specifically, the James Bond movie theme) that the subject described as orgasmic. Both of these last phenomena are quite strange indeed. As if our brains weren't klugey enough in dealing with duality, crossmodality, and synesthesia, we also now can see how our brains are wired to invite our emotions and memories and other aspects of higher cognitive processes into interpreting our outer world. How we invite the emotions and memories into the game of perception is incredibly revealing to how our brains make sense of the world.

# 16 CONNECTOMES

*How Crossmodal Interactions Work in the Brain*

*"For the sense of smell, almost more than any other, has the power to recall memories and it is a pity that we use it so little."*
—Rachel Carson, ecologist and author

The single-celled organisms that dominate our planet can sense their environment in bits and pieces. Although these organisms are simple compared with the typical vertebrate, they are as fully evolved as any other living thing on the planet. And although their biology cannot be considered perfect (evolution does not strive for perfection), these single-celled organisms have found solutions to the environmental challenges they have faced during their evolution (evolution does strive for solutions). Some single-celled organisms will use very simple mechanisms to sense their environment, and that is good enough for them to survive and pass on the essence of their species (their genomes) to populations of the next generation. Most of the sensing done by single-celled organisms concerns counting how many of their conspecifics are near (quorum sensing), sensing other organisms that they should defend themselves from or consume or exchange DNA with, and other sensing operations that are simple but essential for survival. The perception of a microbe to its outer world, then, is fairly limited and one-dimensional in most microbes. Getting a signal that a predator is near may be the only sense a microbe needs, and in this way many microbes have a one-dimensional perception of the outer world.

As organisms became more and more complex, sensing the environment in different ways became a widely used and successful strategy in evolution. Once an organism has adopted several ways of sensing the environment, the senses can stay independent and send information to the brain without reference to other outer-world stimuli. Or, alternatively, the different senses can evolve to be connected in some way to one another for any number of reasons. Sometimes the connections are made because of structural constraints. If two different senses occupy neural real estate near each other, then connections between them might be made randomly. In other cases where crossmodality of senses offers a successful solution to an environmental challenge, selection may be involved. I have emphasized crossmodality so far because higher-order perception is not possible without the senses cooperating with one another even if they are a team of rivals. So, further discussion of crossmodality and how it affects the biology and behavior and evolutionary potential of organisms becomes an important endeavor in understanding perception.

Combining information from multiple senses is not a uniquely human or even uniquely primate capability. Some mammals combine information from different sensory inputs when the information is perceived to share a link to a larger or overarching sensory item. So, for instance, a high-pitched sound might be coprocessed with light colors because these separate sensory inputs might regularly be associated with a predator or a prey item. There are two possible ways that this can happen, and both might be involved at the same time. An organism may do this, first, because the sensory signals reinforce each other and, second, because they might be processed in similar manners. In the first case, the different sensory signals act like ratchets. When one sense steps up sensory acumen, the other does, too, and so on. In the second case, one sense simply hitchhikes with the other. In addition, it is clear from animal behavior studies that nonprimate mammals can actually formulate very complex perceptions of environmental sensory input, especially when another species is involved (taking us all the way back to "I run away from that," "I eat that," and "I mate with that").

Although little work on this phenomenon in nonprimate mammals has been accomplished at the level of brain structure and physiology, behavioral studies that are akin to the psychological approaches used in human cognitive biology all point to the possibility of crossmodality in many animals. Such crossmodality has an adaptive story behind it because organisms, especially ones that have prominent places in ecosystems, such as mammals, might have a selective advantage if they can rapidly and accurately process and categorize things in their environment. Crossmodality goes a long way to this end. Looking at a few examples will demonstrate the plausibility of this way of thinking about nonhuman primates and other mammals and connections that are made between the senses.

Many of the examples are from carnivores such as cats, dogs, and seals, because these animals are social in nature and observing them has a rich payback. In one example, a naive (meaning that the subject has not previously been tested) captive female sea lion named Rio was given a set of what is called match-to-do tests. The results indicated that Rio could establish crossmodal matching of the visual letters and numbers with sounds.

In match-to-do tests, a behavioral tool called conditioning by reinforcement is used. This approach simply conditions an animal to respond to a particular signal as a result of some reward when the animal gets it right. Two sets of visual classes are established, each with several possible members (fig. 16.1). Rio was conditioned to associate specific sounds with specific items in one class of figures and the same sound with specific items in the other class of figures. She was then trained to classify each of the ten individual figures into their proper classes. So, Rio could associate A with the alphabet class and 4 with the number class. Her conditioning was dependent on the prize she got for getting the classification right and a particular pitch. Rio learned this system quite easily and quickly, and she accurately learned to associate high pitch with letters and low pitch with numbers. She was then introduced to six contrasting tones that were paired (Ring/Siren, Sweep/White, and Pulse/Tone) and different from the original high and low pitch. The six sounds were paired to be like the high-pitch and the low-pitch contrasts. Rio was then "asked" to associate a number or letter with the new sounds. Of the six sounds, Rio consistently associated three of them

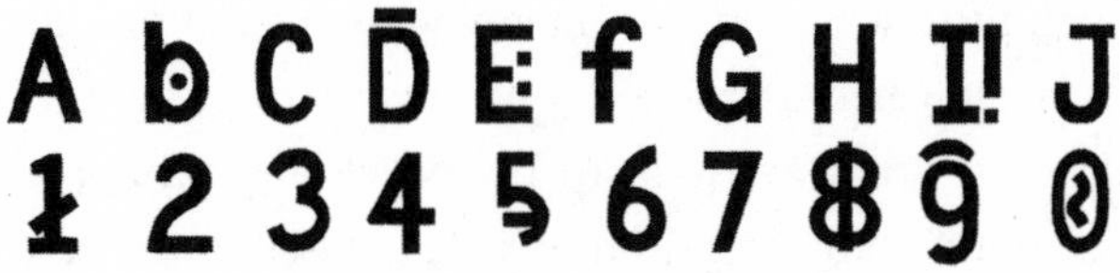

# A b C D̄ Ë f G H I̤ J
# 1̷ 2 3 4 5̧ 6 7 8̷ 9̂ 0̸

Figure 16.1. The figures on the top line are one class of stimuli (alphabet), and the figures on the bottom are the second class (number).

(Ring, Sweep, and Pulse) with letters and three of them (Siren, White, and Tone) to numbers. Rio apparently established crossmodal contexts for the sounds with their corresponding visual results.

It has also been suggested that lions, another carnivore species, use crossmodal responses to identify individuals in captivity and in the wild. The researchers who examined this phenomenon used an approach called expectancy violations theory (EVT) for testing African lions in the wild. This approach measures the response of individuals to artificially applied, unanticipated departures from normal social responses. Again, carnivores are a good group to do this with, because they are social animals in general. In this test, lion test subjects heard loud roars from behind a screen where another lion had originally been. The test was designed to induce a reaction from the test lions, indicating recognition of another individual. Lions recognize each other using both sight and sound, but researchers asked, Is the recognition crossmodal? The researchers hypothesized that if no crossmodality was involved in individual recognition, a lion test subject would respond in the same way to a fully visible lion roaring as to a fake, concealed roar. After testing several lions, researchers noted a difference in response: the lions reacted oddly to the incongruent signals and socially normally to the visually connected roars. These results suggest that there is a crossmodal audiovisual connection in individual recognition of these social animals.

Until recently primates have been studied for crossmodal responses because they can be reared at facilities and can be brought into the lab to be tested. This trend is changing dramatically with our recognition of the ethical dimensions of keeping these animals in captivity for research. Nevertheless, macaques and chimpanzees have been the subject of significant research in crossmodality. In a somewhat stunning experiment in 2011, Vera Ludwig, Ikuma Adachi, and Tetsuro Matsuzawa showed that chimpanzees had the same crossmodal connection of vision and auditory senses that humans have. The tests they used challenged chimpanzees in timed manual discrimination response studies. These studies are compelling to watch, because the chimpanzee is placed before a touch-screen computer and stimulated by both sounds and sights and then asked to respond by touching the screen in the proper place. After many tests with several subjects, it was obvious that the chimps associate white with high pitch and black with low pitch. Sound familiar? It should, because it is pretty much the same response that adult and infant humans have to high pitch and light images.

How did this experiment work? The researchers briefly showed a chimp a small white or black square on the computer screen. After a white *or* black square flashed on the screen, two larger squares (one white and one black) were shown in the upper half of the screen and the chimp was asked to pick the one that flashed before. The researchers recorded how quickly and accurately the chimp responded, and they used these results as a metric in later comparisons with variations of the experiment. The chimps became very used to the test and could do it as quickly as a human. Watching the videos of these tests, one can't help but root for the chimp to be faster at the task than humans. The experiment began once the chimps got used to the black-or-white identification task. As the squares were flashed, high- or low-pitched tones were sounded as a background noise, and the chimp's response was recorded for accuracy and speed. If the high pitch was broadcast along with a white square, the chimp more quickly identified the square as white. Conversely, if the high pitch was broadcast with a black square, the chimps slowed down a bit to get the identity of the square correct.

From these experiments one might argue that the crossmodality connecting light colors with high pitch existed in the common ancestor of chimps and humans and perhaps even further back in the primate lineage. Warning! The following may be a Panglossian explanation for how the crossmodal connection of light with high pitch may have evolved, but it's still fun to think about. (Remember that a Panglossian explanation is one that looks for an adaptive reason in everything.) Some researchers have proposed that the phenomenon of light images being mapped onto high pitch emanates from our lower primate ancestors. The Panglossian argument is based on the notion that there is an adaptive need to process external information fast. The adaptive argument for this con-

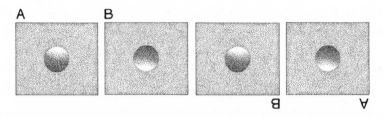

Figure 16.2. The light-from-above effect. In two leftmost panels, the visual system processes A as a convex image and B as a concave image. The visual system does this because of the light-from-above effect. To the right, the left two panels have been rotated by 180 degrees. The same figure that was convex (right side up A) is now concave (upside down A) and vice versa.

nection is also dependent on a trick the visual systems play on us called the light-from-above effect (fig. 16.2). The illustration here shows how the perception of lighting from above drastically effects what we see. The concavity or convexity of the figure is determined by where light is coming from or where we sense that light is coming from. Humans and chimpanzees both assume that light always comes from above—hence, the difference in the perception of convexity and concavity.

The Panglossian argument is as follows. We associate light from above with high-pitched sounds because other organisms that we routinely see above us are usually smaller than we are. For example, if you are sitting outside and are wondering what living creatures are above you without looking, you probably don't even consider that a hyena is up there. Rather, your options are birds, squirrels, or (if you are sitting in Central Park in New York City), the occasional red-tailed hawk. Most of these organisms are indeed smaller than we are and make squeaky, high-pitched sounds. And most of the organisms above us that are squeaky are lit better than organisms that aren't. The adaptive story has one more step to it. Organisms need to know right away if they need to flee ("I run away from that") because survival often depends on recognizing something dangerous in fractions of seconds. Usually organisms don't run away from creatures smaller than themselves and therefore from ones making high-pitched sounds. Conversely, a low-pitched sound means something completely different.

It took me about five hundred words to describe this adaptive just-so story, when a simpler explanation exists. Charles Spence and Ophelia

Deroy suggest that the correct correspondences are actually light with size and size with high pitch. Because crossmodal connections are transitive, once the light-size and size-pitch correspondences are learned, this leads to the light-pitch connection. Instead of there being an innate, direct crossmodal connection of light with pitch in the brain, the correspondence might simply be a learned effect. This alternative scenario also suggests that the environmental context of potential crossmodality needs to be examined before we can make claims of innate, direct crossmodality and synesthesia. The Panglossian explanation can trick us much too easily. But if crossmodality is implemented by the wiring of the brain, can we see differences in neural connections in the brain among organisms? And what then makes things different at the neural level as vertebrate lineages diverge?

Some parts of the brain expand, and others commit to more and more connections to other parts of the brain in the evolutionary process. The case of the primate brain needing to be folded into rolls (gyri and sulci) is a good demonstration of different parts of the brain getting larger in our lineage. In fact, it is well known that there was a huge jump in brain volume in the common ancestor of *Homo erectus, H. neanderthalensis,* and *H. sapiens.* Chimps and some of our older ancestors related to australopithecines and the very early diverging genus *Homo* species had brain sizes all less than 43 cubic inches, with chimps at 18 cubic inches. By contrast, the brain of *H. neanderthalensis,* like that of *H. sapiens,* was on the order of 92 cubic inches (with the average Neanderthal brain being a little larger than the average *H. sapiens* brain). Unfortunately, there is no way to delve into what kind of neural connections were in the ancestors we had with these genus *Homo* species. Unlike Leborgne's brain, which is preserved in formalin, and Molaison's brain, which was sectioned into thousands of thin slices soon after his death, we simply do not have preserved brains or images of the brains of these long-extinct *Homo* species. We do, however, have information on the neural connections of chimpanzees, other primates, and mammals. What can they tell us about the transition into human perception and perhaps even the development of consciousness in our lineage?

The short answer is that only some of the connections are there. The longer answer requires consideration of the connectome. This relatively new brain term refers to the mapping of the axonal connections by means of synapses in the brain and the tendency of modern biologists to append the suffix "-ome" to any approaches that are big data in nature. One way to do this is to simply look at the cellular structure of the brain, as with Henry Molaison's brain described in Chapter 10. This approach has been used in model organisms with manageable brains such as the nematode (*C. elegans*). The approach requires that the brain (or prain—see Chapter 2) of this nematode is sliced into hundreds of very thin sections that are then viewed sequentially in a transmission electron microscope. Each section is photographed at a magnification so that every neural cell can be identified. The photographs are then stacked using computer imaging, and the various neural cells can be tracked and plotted. When this is accomplished over the entire brain, the connections of all of the visible nerve cells can be traced. In a tour-de-force study in 1986, researchers were able to determine that the connectome of the nematode consists of 279 cells. That turns out to be about one-fourth of the total number of cells in this tiny wormlike organism of about 1,000 cells. In addition, there are a precise 6,393 synapses connecting these 279 cells, with 1,410 connections of these neural cells to muscles. Yet this very simple worm can sense light, tactile input, and odors, and can taste. There is no evidence, though, that the nematode has any cross-modal sensory capacity.

This approach is as tedious as it sounds but incredibly rich in information. Although it can give cell-to-cell connectome information, other approaches can give a macroscale picture of the connectome, and the methods used to uncover these more macroscale connectomes include MRI and DTI (see Chapter 15). By combining information from hundreds of studies into a rich database, researchers have attempted to construct neural connection networks. The major analytical tool they use to build connectomes of different organisms concerns the use of a branch of mathematics called graph theory. Basically, each picture of the nervous system—whether it is an actual microscopy picture or simply information about connectivity based on other observations—can be represented in a graph that is then used to analyze and interpret the data.

A graph is simply a mathematical description of the network of connections from one region of the brain to another. The regions that are connected are called nodes, and the resulting graph is intensely rich in information. To combine graphs from different sources is not trivial, so graph theory approaches are needed to make sense of the overwhelming amount of data (big data) involved. Graph theory leads to three very interesting aspects of neural networks: hubs, clubs, and communities. A hub refers to a highly connected node or neuron, a club to a collection of neuronal connections that are much higher than the background level of connections, and a community to an interconnected set of neurons. All three aspects of graphs can be used to give a detailed picture of connections in the brains of different organisms and lead to some universals about connectomes.

Graph theory analysis of the connectome shows that, no matter the species, communities of connections are localized to specific parts of the brain. This pattern means that clustering of neurons most commonly occurs in localized, spatially restricted regions in most mammal brains. Some connections reach out from these communities, but the overall preponderance of connections occurs among close or neighboring neural cells. Moreover, the communities can be linked to specific functions, such as the sensory regions of the brain discussed previously (see Chapter 15). There is also a conservation of these regions across relatively closely related species, such that, for instance, mice, monkeys, and humans show some very similar limbic and cognitive networks. If one branches farther out on the tree of life—to insects or even to fish—these similarities either disappear or are very hard to pin down.

And what all of this means is that, as the brain evolved to become larger and larger in the primate lineage, the larger mass resulted in more tightly wired communities of neurons, which in turn led to the potential for specialized functions of the localized communities. Martijn van den Heuvel, Edward Bullmore, and Olaf Sporns suggest in their review of connectomes that the strong modularity of structures that are observed in the connectome have led to the localization of cortical regionalization of sensory and motor information. In addition, they suggest that as this hub, club, and community organization of the connectome emerged, it also led to increased lateralization of the brain and the evolution of specialized neural functions like language. Indeed, the acquisition of higher cognitive functions that make us the species into which we have evolved more than likely would not have occurred without these general rules of clustering of neural cell connections in the connectome. And of relevance to the crossmodal effects and synesthesia introduced in Chapter

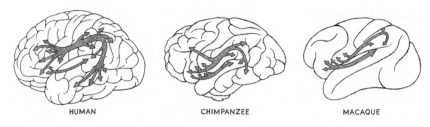

HUMAN          CHIMPANZEE          MACAQUE

Figure 16.3. Connectomes of the human, chimpanzee, and macaque.

15, this general rule about connectomes favored short neural pathways over longer ones as well as more communication between communities that results in crossmodal effects.

But the story isn't finished, because it doesn't end with the mere observation of similarities. There are differences, too, between humans and other mammals and especially other primates. The most relevant to the discussion are those among the lower primate macaque; our closest living relative, the chimpanzee; and us (fig. 16.3). The surprising result in comparing the connectomes of these primates with ours is that in many ways they can offer us some hypotheses concerning higher cognitive and behavioral aspects of our evolutionary history.

The connections in the brain that have undergone evolutionary change can be determined using the various brain imaging techniques discussed previously (see Chapter 15). The two major tissues of the brain where connections are critical—gray and white matter—have distinct differences among primates. White matter is white in appearance and made up primarily of nerve cells called axons that are threaded through this part of the brain and act as conduits to the gray matter of the brain. Gray matter is pinkish gray. It is more complex than white matter in that it contains a different kind of nerve cell called a dendrite and also contains the ends of the axons that run in from the white matter. The ends of the axons form synapses with the dendrites. The gray matter is generally found in the outer layers of the brain, and the white matter is situated more toward the inside of the brain.

Connectome tracts are mostly in the white matter, and indeed this is where the most important tracts of neural tissue are involved in more intricate connections. Diffusion tensor imaging is an especially good technique for deciphering the connectivity of tracts in white matter. Humans, in comparison to other primates and even in comparison to other

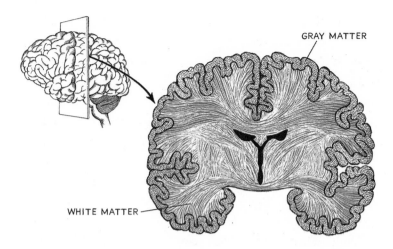

GRAY MATTER

WHITE MATTER

Figure 16.4. White and gray matter in a cross-section of the brain.

mammals, actually differ in the density of white matter, which means that there are also probably differences in the density of connections. Specifically, humans have a higher ratio of white to gray matter in their brains, and this occurs most conspicuously in the prefrontal cortex (Fig. 16.4).

As for connectivity, researchers have determined that only one-fourth of the landmark connections among primates are common to monkeys, apes, and humans. There are also specific areas of human brains that have inordinate numbers of connections, and these correlate with the specialization of function for the region. Good examples of this trend are regions such as Broca's and Wernicke's that are involved in such uniquely human capacities as language. Examination of connections using DTI reveals that connections in several parts of the primate brain (mostly areas in the outer middle part of the brain) are conserved from monkeys to apes to humans. But other parts of the prefrontal cortex and the inferior parietal section (a region just posterior to the prefrontal cortex) of the brain show very interesting and more complex divergence in humans.

As lower primates diverged from monkeys and monkeys from the great apes, the brain enlarged and connections in the more derived ape brains (like ours) were added to. The mirror neuron systems in primates are a good example of this trend. Mirror effects were first shown in macaques, and both chimps and humans also show the phenomenon. Brain imaging techniques such as DTI can uncover the neural connections of the mirror system, and researchers have examined the fine detail of such connections. Diffusion tensor imaging has shown that mirror neuron systems have three levels of organization in primates. Macaques, chimps, and humans have mirror neuron connections in the frontal temporal part of the brain, and the connections extend into the frontal parietal region of the brain by means of very specific neural cells. But this region is where the mirror neuron pathway stops in macaques. In humans and chimps, the mirror connections extend into the inferior temporal cortex. But this is where chimpanzee mirror connections stop. In humans alone, the mirror connections extend into the superior parietal cortex. Erin Hecht and her colleagues have constructed a model to explain these differences and suggest that "differences in mirror system connectivity and responsiveness with species differences in behavior, including adaptations for imitation and social learning of tool use," might have evolved in this way.

But here is another Panglossian moment. It seems that a major goal of the comparison of macaque, chimp, and human brain size and brain connections is to explain the broad difference in behavior of the species involved. There is no doubt that the source of the huge difference we see among ourselves and other primate species is indeed the brain. Yet we must be careful about allowing Dr. Pangloss to make a house call and leap to the conclusion without evidence that these different mirror system brain connections led to an evolutionary jump like tool making. On the other hand, such a suggestion is a wonderful hypothesis for testing.

# 17 FACES AND HALLUCINATIONS

*Facial Recognition and Hallucinations as Subjects in Higher Perception*

*"I am having a hallucination now, I don't need drugs for that."*
—Thomas Pynchon, writer

Macaques are good at recognizing faces and depend on facial recognition for much of their social organization. Readers of this book, unless they have a cognitive mix-up called prosopagnosia, should know how important facial recognition is in our species. There are two types of prosopagnosia, one caused by injury and the other congenital. In both forms the fusiform gyrus, a structure deep in the visual processing region of the temporal lobe, is affected, and hence this region of the brain appears to be responsible for processing visual information about faces. Indeed, macaques' facial recognition networks can be mapped to this region, too. Although it appears that human facial recognition networks are located more toward the back of the brain (ventrally) in this area than in macaques, the way the monkey and human brains develop may produce this seemingly clear locational difference. And yet, the connectivity of facial recognition neurons shows clear differences between human and macaque. Connectivity studies in humans have indicated that there are two face-processing streams of neurons in this area of the brain that work kind of like the visual what-and-where streams dis-

Figure 17.1. The T-shaped face in renderings of a Giuseppe Arcimboldo painting of fruit and vegetables, right side up and upside down. The T shape is easily
recognizable in the upside-down orientation (right), and hence we interpret this
painting as a human face. Even when the painting is right side up (left), some
people detect it as a face because they recognize the upside-down T, and the
information then is shunted into the face recognition visual network.

cussed in Chapter 5. Like the visual what-and-where streams, for the
face recognition streams one is dorsal and one is ventral, and they have
only weak connectivity between the two. Macaques have the dorsal
stream and even have rich interconnectivity in this stream, like humans.
But the human lineage has evolved the second stream independently.
These results suggest a major difference in how faces are recognized in
macaques and humans. But what about chimpanzees, which are a much
closer relative to humans than are macaques?

Jessica Taubert and Lisa Parr have examined how chimps respond
to faces. To understand their approach, it is important to recognize that
human facial recognition brain areas respond more directly to faces than
to such other items as shoes or chairs. All faces have what is called a
T-shaped pattern (fig. 17.1). The two eyes form the bar of the T, and
the stem of the T represents the nose and mouth. Recognition of this T
shape is what Taubert and Parr call first-order information about facial
recognition. Recognition of the T shape is a first step that allows the
information to be pushed further down the face-recognition neural pathway into what are called second-order pathways. Taubert and Parr first

asked whether the chimp response to faces is based on the first-order information of the T shape or is more complicated, using what are called Mooney objects. These are images of faces or other objects that are in black and white with low contrast and little information in them with respect to typical facial features.

Taubert and Parr created Mooney objects of chimp faces, human hands, and inanimate objects such as shoes. The trick is to create a series of Mooneys with graded black and white tones of varying contrast. For humans, when contrast is optimized on a Mooney series of figures, the second-order pathway of the human face recognition system is diminished, and the only thing we use for recognition of the image is the first-order information. The second-order information is what we use to recognize individuals, so what happens in the case of a high-contrast Mooney face is that we can recognize it as a face, but we cannot recognize the individual.

Chimps can easily identify faces as faces. They can easily discriminate between human hands and shoes versus faces, too. And like humans, chimps fall back on first-order information when the contrast is ratcheted up and lack the ability to identify individuals in these cases. The results of these experiments suggest that chimps share this level of organization for facial recognition and have been part of the extra step of two visual facial recognition streams of neurons, which is a totally reasonable conclusion, considering that we shared a common ancestor with chimpanzees five to seven million years ago. What this also means is that those genus *Homo* species discussed previously (see Chapter 16) also probably had this more complicated visual facial recognition wiring.

Why is facial recognition so important? Many biologists argue that recognizing faces is an integral part of socialized behavior in humans and other primates. Visual recognition of faces and other objects is indeed important in an adaptive context. It is difficult to deny Dr. Pangloss a seat at the table for this one. As discussed in Chapter 10, certain brain injuries that result in interesting facial recognition phenomena, such as Capgras syndrome ("I know you look like my mom, but I am not reacting to you like you are my mom, so you must be an imposter"), and the split-brain "Mike or Me" effect are good examples of how the process works with faces. But what about other objects that mix up the sense of vision? Witness the viral image of "dog rear Jesus" that in one week's time received nearly two hundred thousand web hits and has since become an Internet classic. Dog rear Jesus is an apparently unretouched photo of a dog's rear end with an anus that resembles the face of a long-haired bearded man. More stunning is that the surrounding fur looks like a body in a robe with outstretched arms. It doesn't take much for the brain (at least mine) to see the Lord and Savior in all of his humble glory.

And indeed, you would have to be a doubting Thomas, a human not familiar with Jesus, or a lying sinner to deny that the dog's rear end resembles a ten-inch-tall replica of the Savior himself. Dog rear Jesus is not the only religious icon we see in strange and seemingly inappropriate places. For one, the Virgin Mary must have had as many fashion looks as Lady Gaga. Her likeness has appeared on everything from the walls of freeway exits to the surfaces of sandwiches. About a decade ago, a piece of a grilled cheese sandwich with her supposed image on it was auctioned off for twenty-eight thousand dollars. These sightings are not simply recent phenomena, because apparitions of the Virgin Mary have been reported ever since the Assumption—many famous and infamous images of that event, and of the Virgin, have been created by artists (including one created with elephant dung as paint). And when you throw in the toasted cheese, the water markings on walls, and assorted other apparitions, she appears in an amazing assortment of guises.

There is actually a word for this phenomenon—pareidolia—and it is a specific form of a neural process called apophenia. Apophenia is the interpretation of random data by the brain as something meaningful. Pareidolia comes from the Greek *para,* meaning "instead of" and *eidōlon* meaning "image." In this context, it means "faulty image." Examples of pareidolia other than imagining religious icons abound. The premise of the famous Rorschach tests is one. These tests involve the interpretation of rather randomly produced ink smears or blots. Psychologists then use the pareidolic interpretations of the ink blots for psychological interpretations. There are also auditory versions of pareidolia. Perhaps the most famous examples are "I buried Paul," supposedly heard at the end of the Beatles' song "Strawberry Fields," and the purported satanic chanting in the backmasked (played backward) playing of Led Zeppelin's "Stairway to Heaven."

The question, then, is why do our visual and auditory senses get so twisted up that we see Jesus in a dog's rear end, recognize the Virgin

Mary on grilled cheese sandwiches, or hear satanic verses when songs are played backward? What is it about the brain that allows for such unholy profanity? Most of the answers lie in the kluginess of how the brain has evolved and how brain cells communicate with one another and with the outer world.

One last aspect of the brain that is in play when we see the Virgin Mary on a piece of toast or Jesus on a dog's rear end is that the brain has more than likely been wired to prefer religious explanations for unusual phenomena. Most evidence for this explanation is anecdotal, but there is some experimental evidence, albeit controversial. The anecdotal evidence comes from evolutionary psychology, the branch of evolutionary biology that used to be called sociobiology, where evolutionary explanations are offered for human behaviors. The evolutionary psychology explanation for human dependence on religion, generally put, is that it offers social cohesion to populations practicing it and that this in turn is an evolutionary advantage for the population. And yes, I should be issuing a Dr. Pangloss warning here, too.

The other purported evidence that religion is perhaps hardwired is the product of some relatively wacky experiments involving the electromagnetic stimulation of specific regions of the brain. These experiments involve the so-called "god helmet," a device developed by Michael Persinger. Stimulating the brain and then observing the resulting behaviors is not a novel approach to the study of brain function. Remember that neurosurgeon Wilder Penfield tickled various regions of the brain during his patients' surgeries and asked them what they felt (as pointed out earlier, the surface of the brain has no pain receptors, so some brain surgery is done with the patient wide awake and conversant). These probes allowed him to map the real estate of the brain responsible for specific sensory and motor functions. The sensory and motor homunculi discussed in Chapter 3 are the results of this important work.

Using the same principle, but in a noninvasive way, Persinger outfitted a helmet with the ability to transmit electromagnetic radiation to highly specific regions of the brain of the wearer. Electromagnetic radiation apparently alters the function of the neurons in the radiated region. Persinger hoped to demonstrate the existence of a specific region of the brain responsible for ecstatic religious feelings. Some subjects report light-headedness and feelings of well-being along with a spiritual feeling when under the influence of the god helmet. Others, like the famous evolutionary biologist Richard Dawkins, report no impact at all. Although the evidence is anecdotal for a god region or a god gene, our social behavior is an oddity in the animal world. Our dependence on religion and spirituality to maintain social cohesion is more than likely the result of rapid cultural evolution. As Dawkins explains, "Religion is about turning untested belief into unshakable truth through the power of institutions and the passage of time." But whether the visions and apparitions many report seeing when experiencing religious ecstasy are real is another story. Perhaps Dawkins's answer to a question from a rather zealously religious attendee at one of his talks provides a clue. The question was: How can Dawkins explain the "fact" that the questioner during prayer actually sees religious icons and walks with Jesus and Mary daily? Dawkins responded that he didn't doubt that the individual and others who do see these apparitions are "seeing" them and, more important, he didn't doubt their sincerity. But as Dawkins added, "Sir, I believe you are hallucinating." This brings to mind one other reference to the image of the Virgin Mary in popular culture, and that is from the famous TV series *The Sopranos*, when wise guy "Paulie Walnuts" Gualtieri momentarily sees the Virgin Mary near a stripper pole in the Bada Bing. His vision was not a case of pareidolia but rather a fleeting hallucination, and these are another story about messiness in our brains that now warrants consideration.

My first reaction to reading Oliver Sacks's book *Hallucinations* was, "Can I get some of what he was on?" Sacks was always starkly frank about his experimentation with mind-altering drugs as a young man. In his very first experience with LSD, he comically relates that after he and a friend dropped a tab of acid they had mail ordered, they waited for it to kick in by listening to the music of Igor Stravinsky, hoping to hear something mind-bending. They quickly realized that the Stravinsky piece sounded the same as it always did and that they had short-armed their acid dose by an order of magnitude. They had taken only enough LSD to cause a cat to trip. In another drug-related story about his experiences, he injected a particularly large dose of morphine and observed the Battle of Agincourt on his dressing-gown sleeve for what he thought was a few minutes but in reality was twelve hours.

Hallucinations can be induced by drugs, as Oliver Sacks's experiences show, but they can also be induced by brain injury, migraine, or dementia. Or they can be symptoms of psychiatric disorders. Hallucinations can present in schizophrenia disorder, a broad-spectrum psychiatric disease, and in Parkinson's disorder and other forms of dementia. Hallucinations can affect or use all of our major senses, including the vestibular or balance system. Four of the big five senses seem to be involved in most hallucinogenic experiences of schizophrenics—vision, hearing, tactile sense, and olfaction. Auditory hallucinations appear to be most commonly experienced by people with the disorder, with visual hallucinations following next. But vision and hearing together seem also to be a rather common category of hallucination for schizophrenics. Because schizophrenia is a broad-spectrum disorder, it is difficult to pin down how the senses are affected by this general kind of psychosis.

Dementia is another group of psychiatric disorders where hallucinations occur. Two disorders—Parkinson's and dementia with Lewy bodies—are grouped under the major category called Lewy body dementia. In the early 1900s, Frederic Lewy, a Berlin-born Jewish neurologist who worked in Alois Alzheimer's lab in Munich before fleeing Nazi Germany to become an American citizen, discovered conspicuous anomalies in cells of the dopamine-producing area of the brain, called the substantia nigra, of patients who had died of dementia. This area of the brain is a darkened region at the tip of the pons section of the brain stem. The abnormal neural cells look very different from unaffected neural cells when stained a certain way and observed under a microscope. The abnormal cells swell and riddle the brain tissue of people with certain kinds of dementia. They appear to fill up with a protein called alpha-synuclein that tends to diminish the number of dopamine-producing neurons in the brain. People suffering from Alzheimer's disease will also have Lewy bodies, but the major locus of pathology for Alzheimer's patients occurs in the hippocampus. In addition, since the Lewy bodies are localized in the substantia nigra, the drastic effects of Alzheimer's disease on the overall structure in the brain do not occur in people with Lewy body dementia, so Alzheimer's is not considered a form of this dementia.

Whether hallucinations are present is one test of three that physicians use to diagnose Lewy body dementia. Self-reported hallucinations are one way to find out whether a person hallucinates, but another, more objective way is to exploit the pareidolia effect just discussed. In one version of the test, subjects are shown a series of blurred pictures of natural scenery for sixty seconds during which they are asked to describe what they are seeing in as much detail as they can. They are asked to point to objects in the pictures as they describe them and are not told whether they are correct while taking the test. Their answers are classified into three categories. The simplest category is the subject's stated inability to recognize the scene or any objects in it. A second response is an accurate description of the scene and the objects in it. The third category includes any illusory or incorrectly identified objects. If a potential illusory object is identified, the individual is asked if the object is in the picture or if the object looks like what was described. Other versions of the test use the same principle of identifying illusory objects, and the degree of pareidolia scored is correlated to the number of illusory items observed by the subject. The use of pareidolia to diagnose Lewy body dementia seems to be a good way to objectively identify the disorder, and it also points to potential study of people with this disorder to pin down the neurophysiological basis of hallucinations.

So, what is the neurological basis of hallucinations? Early work on this question involved examining people who experienced hallucinations and then attempted to synthesize an overarching theory about them. The first step in getting at the source of hallucinations was to define them. In the 1930s, both hallucinations and illusions were thought to be part of the same thing—visual anomalies. Psychiatrists then decided to tease apart illusion from hallucination, and this better defined what these visual experiences consisted of. At this time, too, there were several theories about the origin of hallucinations. One suggested that they were visual anomalies caused by problems in transmitting or interpreting the signals that originated in the retina. Hence, eye injuries or diseases were the source of hallucinations. This ocular theory of hallucinations appeared reasonable because patients with eye disorders seemed to hallucinate more than individuals without eye problems, and people with eye disorders who hallucinated were cured of their hallucinations when their eye disorders were cured. But the idea that eye disorders alone caused hallucinations was rejected, and the medical state of the eye, although not excluded as part of the story, was rejected as the sole explanation.

One particular kind of hallucination that advanced thinking about the nature of these anomalous sensory experiences involved a syndrome about which Oliver Sacks wrote at length: Charles Bonnet syndrome. This syndrome affects people who have lost sight as a result of old age. These older blind people have vivid and complicated, usually pleasant, visual hallucinations, even though they have lost the sense of sight, and what's more, they freely admit that they are hallucinating. To some physicians, Charles Bonnet syndrome was the clearest example of why the eye should be decoupled from hallucinations. But to others, the eye was kept in the picture as part of the explanation. Nonetheless, Charles Bonnet syndrome became the model system for studying the origin of hallucinations. It appears that if you can produce an explanation that also works for Charles Bonnet syndrome, you have hit the jackpot in the explaining the hallucinations sweepstakes.

Another important aspect of research on Charles Bonnet syndrome is that it has forced neurobiologists who study it and other syndromes with hallucinations to be very picky about what they are calling hallucinations. Researchers have realized, as with schizophrenia, that hallucinations fill a spectrum of phenomena. A comparison of two kinds—zoopsia and Lilliputian hallucinations—shows the potential for differences among hallucinations. Zoopsia hallucinations involve animals, and they progress in a well-defined way. Lilliputian hallucinations involve tiny people much like Sacks's Agincourt reenactment, and they, too, have well-defined characteristics.

When different kinds of hallucinations are analyzed as separate phenomena, it becomes evident that different neural pathways or complexes of neurons cause the different kinds of hallucinations. Visual hallucinations are one of the more common and easy to follow kinds and have been the object of several fMRI studies. Other types of sensory hallucinations have also been evaluated. For example, in a study of subjects with schizophrenia, the sensory category or categories of more than five hundred hallucinatory events were recorded. Auditory hallucinations were the most common, followed by auditory and visual hallucinations (fig. 17.2).

Researchers Dominic ffytche and Paul Allen have been at the forefront of this neurovisualization work. Allen and his colleagues reviewed the large literature on hallucinations and developed a model for auditory hallucinations. The model involves the hyperactivity of the areas of the auditory cortex that process the primary information coming into the

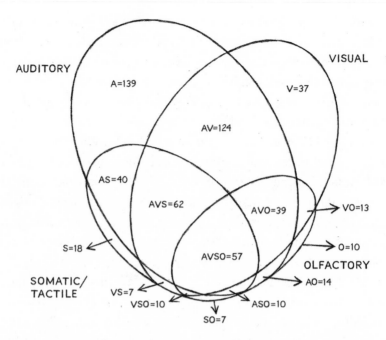

Figure 17.2. Venn diagram showing the relative occurrence of different kinds of hallucinations and the overlap or combination of sensory kinds of hallucinations.

brain from the ears. It also includes the idea that altered connectivity exists between these primary auditory cortical regions and language processing in the inferior frontal cortex. The model further requires that weakened control of the entire system by other regions of the brain responsible for auditory processing should exist. While establishing the preconceived idea that the connectivity in the auditory neural system should be altered in people who hallucinate, how the altered activity happens and how it affects our sensory perception is not known.

ffytche and his colleagues have developed a "taxonomy" of hallucination phenomena, carefully separating different kinds of hallucinations as they analyze their functional MRI data. The approach involves putting a hallucinating subject into an MRI machine and asking when hallucinations start and end. The subject is also asked to describe the hallucination so that they can categorize it. Using these approaches, ffytche and colleagues have established that brain activity associated with hallucinations is highly localized. When comparing the activation patterns of people having different kinds of hallucinations, they observe subtle differences in the neurons activated.

The visual pathways of the human brain are pretty well worked out. We know color is processed in the visual cortex, where information that is used to identify faces is processed, and even where specific parts of the face are processed. So, when the subject being examined with fMRI was having a very colorful hallucination, we know that the active region was the part of the visual cortex responsible for color interpretation. The correspondence of the regions of the visual cortex with the specific types of hallucinatory images was striking.

Other evidence for a localized nature of hallucinations in the brain comes from comparing people undergoing auditory and visual hallucinations. Functional MRI studies show clearly that these two major kinds of hallucinations activate different regions of the brain. Audio hallucinations often involve speaking, and so the relevant brain regions for speech, such as Wernicke's and Broca's areas, show activation during auditory speech-related phenomena.

This all makes good sense, because these brain regions control many aspects of vision, speech, and speech comprehension. So, are people who hallucinate really hearing and seeing? Their brain activation patterns suggest that they are hearing and seeing something, but is it really sight and sound? Remember mirror neurons, where specific parts of the brain that control sensory input can be activated simply by watching someone else do something? Some researchers suggest that people who hallucinate actually do feel that the sounds they "hear" are coming from the outer world or, in other words, misattribute hallucinated voices and sounds to real external stimulus. If this is the case, then people who hallucinate have tricked themselves into something akin to being able to be tickled by themselves. And indeed, Sarah-Jayne Blakemore and her colleagues have shown that normal individuals cannot tickle themselves efficiently because self-generated tactile stimulation is dampened as a result of self-awareness of where the stimulus comes from. But when people who experience auditory hallucinations are tested, they appear very capable of tickling themselves as efficiently as if someone else was tickling them. Although the exact mechanism for this tickle thyself phenomenon is unknown, it is thought that there might be a disconnect or altered connection between how normal people monitor the initiation of an act and the regions of the body that perceive the results of the action. People who regularly experience hallucinations are not good at making the connection between their action and the sensory information from that action. They are also not good at keeping time or placing actions into a temporal framework, and this contributes to the disconnect that allows them in turn to experience their hallucinations as not coming from self.

When Oliver Sacks and his friend dropped the weak dose of LSD, the chemical technically known as lysergic acid diethylamide went into the stomach, was absorbed by the digestive tract, and then passed into the blood, where the small LSD molecule was transported to the brain. This compound is structurally very similar to serotonin, a neurotransmitter in the brain, and so one of the many receptors in the brain that reacts to serotonin in particular is terribly confused by the presence of the LSD. The confusion caused by LSD's interaction with this serotonin receptor results in altered connectivity of specific regions of the brain similar to the messed-up connectivity of people who hallucinate from brain injuries and neural disorder. But the dose Sacks took was so low that only a small proportion of the serotonin receptors were affected, and Sacks was not affected by the attempt to trip out. He did eventually get the dosage right, and when he did, the impact on his brain chemistry was very different.

It has been nearly fifty years since Oliver Sacks first tried LSD, and only now are scientists getting closer to cracking open what the brain on LSD is all about. In a novel experiment performed by Robin Carhart-Harris and his colleagues, tripping subjects were examined with three modern brain imaging technologies, including fMRI. The brain on LSD shows altered activity in expected brain regions such as the visual cortex. People who are tripping can detail the intensity of their hallucinatory experience, and the connectivity patterns before and during slight hallucinatory experiences and extreme hallucinatory experiences can be quantified. In this case, people who are tripping their brains out show expanded connectivity of their visual cortex with other regions of the brain. More important, these studies show that connections of the primary visual cortex to two areas of the brain not considered part of the visual pathway (the parahippocampus and the retrosplenial cortex) are significantly diminished in people on LSD. As the connections to these regions of the brain get weaker and weaker, subjects experience more and more dissolution of self and loss of ego. These results suggest that LSD has the same self-tickling impact on the brain as for other sources of hallucinations, by weakening the connection to the visual phenomena and the sense of where the stimulus is coming from.

It is also obvious that memories and emotions intrude on the neural processes of people who are hallucinating. There is a certain amount of overstimulation of the sensory areas of the brain responsible for auditory and visual hallucinations that initiates hallucinations, but this is augmented by unusual suggestive input from other regions of the brain that are involved in higher-level cortical processing. In one study, a person experiencing visual hallucinations was examined for unusual connectivity in the brain. The study showed hyperconnectivity between the visual cortex area and the amygdala, which is embedded in the limbic system and is heavily involved in emotional response. Although this study involved only a single individual, it is highly suggestive that hyperconnectivity might be a factor in directing or suggesting emotional content for hallucinations.

Although Oliver Sacks engaged in some extreme drug use during his life, he was always careful not to encourage it in others, and he also pointed out that his intellectual development and knowledge of the mind were very dependent on that phase of his life. He stated clearly that drugs "taught me what the mind is capable of." Sacks was one of the most influential researchers of the past century in brain science. His treatise on hallucinations is a fascinating tour of the mind and how our senses are pirated away by hallucinations to create alternative forms of perception and in turn alternate realities. Hallucinations taught him several lessons. First, because he was an astute neurobiologist, he noted correlations of hallucinations with neural structural anomalies in people. He commented, "The phenomenology of hallucinations often points to the brain structures and mechanisms involved and can therefore, potentially, provide more direct insight into the workings of the brain." He recognized the importance of hallucinations in how we as a species have developed socially and culturally. And he finally recognized how our perception of the world is needed for us to be conscious beings. Sacks also had an acute interest in other aspects of the brain as sources of information, leading us to a better understanding of our existence, which is how we process and mold our senses, and ultimately perceptions of the outer world, into words, language, art, and music that make us a stunningly unique organism on this planet.

# 18 BOB DYLAN'S NOBEL

*Language, Literacy, and How the Senses Interact to Produce Literature*

"*A poem is a naked person. . . . Some people say that I am a poet.*"
—Bob Dylan, musician and poet

Communicating by symbols (written, sung, spoken, or tactile) is not uniquely human. Chimpanzees and gorillas can use sign language and symbols to communicate with humans. Other animals have unique communication capacities, too. But no other species on this planet has developed communication with pictures, sounds, words, and symbols as intricately as *Homo sapiens*. We also are uniquely capable of developing genealogies of different languages and of creating fictional languages for recreation. Dothraki (from HBO's hit series *Game of Thrones*, fig. 18.1) and Na'vi (from James Cameron's movie *Avatar*) are two recent examples of this well-respected and scholarly endeavor. The Language Creation Society encourages the creation of novel, or constructed, languages (also known as conlang to those in the know). Some online translators even include in their options for translation such languages as Dothraki, Na'vi, Klingon, Elvish (Quenya and Sindarin), Dwarvish (Khuzdul), Entish, and Black Speech. Although such endeavors may seem a bit extreme, by following the genesis of a fictional language, conlangers open up new ways to learn about how our brains work while on language. As

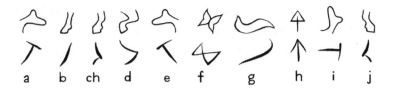

Figure 18.1. Part of the mythical Dothraki alphabet.

the Dothraki might say, "Atthirar nesolat lekh, shafka nesat rhaesheser" (loosely translated as "If you learn language, you know the world").

For more than three centuries humans have also tried to take the elements of different languages and decipher how they are related. Think about this. New languages arise from old ones. They have ancestral characteristics that allow linguists to attempt to reconstruct how certain languages actually evolved. You might be asking how many languages are there, or how many languages have there ever been? The first question is answered from studies accomplished by anthropologists and linguists. Today nearly seven thousand distinct languages are spoken on our planet, but more than half of these will most likely be lost within the next twenty years.

The following is a bald-faced example of a back-of-the-envelope calculation and is a very rough estimation of the answer to the second question, the number of languages that have ever existed. The rate of loss of languages in the next twenty years will be much higher than in the past as a result of globalization, and rates before this century work out on the back of the envelope to about two a year. If we accept that languages go back at least ten thousand years, then probably twenty thousand or so languages have existed on this planet, with 90 percent of them having gone extinct. At this rate, by 2100 probably less than 1 percent of the languages that have ever existed on this planet will be spoken. In other words, 99 percent of all languages that have ever existed on Earth will have gone extinct by the end of this century, and other than a few made-up conlangs, there will be no replenishment with new ones. Not as bad as the extinction rate of species on the planet (99.9 percent of all species that have lived on Earth have gone extinct), but species have been around for billions of years.

The different elements of language such as syntax and vocabulary can be used to figure out which languages are most closely related to one another. So, for instance, Spanish and Italian are very closely related because they share many elements of vocabulary and syntax. But they are both quite different from Tongan, which in its turn is more similar to Samoan than either of those Pacific languages is to Italian or Spanish. Using this general approach, humans have dissected language to do something incredibly unique with spoken words—to understand the genesis of how the varied ways we communicate with one another arose.

To describe the nuances of linguistics and the particulars of language would be a daunting task far beyond the scope of this book. My goal here is to show that language is connected to the senses (easy) and that linguistic capacity, as well as our capacity for music and art, leads our brains to do something unique in the animal world (a little harder) and how this connects us to the world through our minds (much, much harder). The senses involved in language processing are primarily auditory and visual. But tactile senses are also used in braille and other means of communicating words using touch. For instance, some of the fictional languages created by the conlang community involve holding hands and speaking through grips and grasps. And in the long run it would not be surprising to me if a society for the creation of smell as a language crops up sooner or later. Certainly there are enough unique odorants that we humans have the potential to smell to provide us with a large enough vocabulary.

There is a lot of controversy over how and when language arose in our lineage as well as controversy about language in our closest relative, *Homo neanderthalensis*. The problem is that language is not observed directly in the fossil record. Instead, we are left with four indirect approaches to understanding the origin of language. First, we can look at the fossil record with respect to the anatomical structures that produce speech such as the voice box or, more technically, the hyoid bone and its location in skeletons of extant species and reconstructed extinct species. Second, we can examine artifacts such as stone tools and other easily preserved items that might be relevant to language. The idea here is that if an archaeological artifact implies symbolic reasoning and communication through such reasoning, then language more than likely existed for the maker of the artifact. Third, we can make some inferences about the origin of language by looking at linguistic patterns in *H. sapiens* and seeing how these compare to communication in chimpanzees. This approach can pinpoint *sapiens* developments but cannot say with impunity that they are uniquely *sapiens* to the exclusion of Neanderthals or other genus *Homo* species. Finally, we can also take information from living species such as genome-level data that have been shown to be involved in speech or language comprehension and use those data for interpretation of past events in the evolution of language.

Let's look at archaeological evidence first. The most compelling archaeological evidence would be some form of preserved writing such as the hieroglyphics of ancient Egypt. This kind of evidence does not exist for either archaic *sapiens* or Neanderthals. Another way to infer the existence of language from archaeological evidence might be to find evidence of the kinds of thought processes used in constructing language. Because symbolic thought is a major component of language, archaeologists have looked for evidence of this aspect of the human mind. Although the use of symbolic thought is not so hard to detect in our modern objects such as art and music, how an archaic human or a Neanderthal might have expressed it is difficult to ascertain. Ritualistic objects like colored shells in necklaces and formalized burials have been observed in both the Neanderthals and *sapiens* archaeological record. These items suggest some sort of symbolic thought. Perhaps the most interesting and, more important, oldest archaeological item to which symbolic thought can be inferred is a small stone with strong and intentional etchings found in an archaic *sapiens* site in the Blombos Cave in South Africa (fig. 18.2). The artifact is one hundred thousand or so years old and is thought to be the first strong evidence of the use of symbolic thought in human communication. On the other hand, there is scant evidence in the archaeological record that suggests Neanderthals could speak or that they had developed language. This doesn't mean that Neanderthals didn't have sign

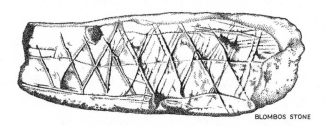

BLOMBOS STONE

Figure 18.2. The Blombos Stone.

language or some other form of very basic language, though, nor that they didn't use symbolic thought. And it is arguable that the archaeological evidence for language in *sapiens* is scant and not much better.

The anatomical arguments about language origins concern the apparatus that *sapiens* uses to make the sounds that result in speech and implement language. The toolbox concerns the placement of the larynx in the throat region of the skeletal system. Specifically, the larynx has dropped to a lowered position in our lineage as opposed to the higher ancestral position seen in chimpanzees and Neanderthals. There is recent evidence that at least one Neanderthal skeleton from the Middle East may have had the larynx lowered, and this suggests that this specimen may have made speechlike sounds similar to ours. Anatomically modern *H. sapiens* appears to have evolved the morphological structures for a working voice box after our species diverged from the common ancestor of *neanderthalensis* and our species about two hundred thousand years ago.

Another example of the approach of focusing on anatomical structures involves reconstructing the brain of extinct specimens. If the parts of the brain involved in speech and language comprehension can be examined and approximate modern human brain wiring for this trait, then we can make some inference about the existence of language in such specimens. So, for example, if we used fully developed and connected Broca's and Wernicke's regions of the brain as benchmarks for speech and language, then a logical approach to determining if Neanderthals spoke (at least in the way we *sapiens* do today) would be to examine Neanderthal brains for such structures. If we could reconstruct Neanderthal brains the same way that Michel Thiebaut de Schotten and his colleagues reconstructed Phineas Gage's brain wiring from images of his skull, discussed in Chapter 6, then we would have an answer. The problem is that the reconstruction of Gage's brain wiring required the researchers to make the assumption that his neural wiring was similar to a reference human brain. This assumption would negate asking the very questions one would want to know about brain regions involved in speech in a Neanderthal. So, the question cannot be approached that way. And because the only tool researchers have for the reconstruction of fossil brains is through endocasts that roughly reconstruct the surface of the brain, paleoanthropologists have yet to precisely reconstruct the brain of a Neanderthal. Hence, there is no way to see what the physical states of Broca's and Wernicke's regions were in this species or in any fossilized archaic human.

What can be done, on the other hand, is to use the brain size and shape from endocasts to make inferences about brain development or anatomy. There is novel evidence that the development of the brains of Neanderthals and *sapiens* were quite different during early critical phases of cognitive development. Such evidence comes from examining the skulls of infant Neanderthals and comparing them to *sapiens* and chimpanzee infant skulls. The evidence from these reconstructed infant brains, when added to earlier studies showing that chimpanzee and *sapiens* brains develop very similarly after the first year of life but not during the first year, indicates that this first year of development in humans and its globular result on the brain has something to do with cognitive differences of chimpanzees, Neanderthals, and *sapiens*.

Philipp Gunz, Simon Neubauer, and their colleagues were able to reconstruct a newborn Neanderthal skull to examine the dynamics of the structure of the developing brain case to make inferences about the shape of the brain that sat in the tiny skull. Both Neanderthal and *sapiens* babies have elongated brain cases at birth. What happens after birth in both species is significantly different, even though both species attain the largest brain of genus *Homo* members. *Homo sapiens* babies develop a more globular brain shape in the critical first year of life, while Neanderthals retain the elongated brain shape. What this means is that *sapiens* brains develop to become bigger in a very different way than Neanderthal brains, and more important, the growth differences occur in a period of brain development that is critical for increased cognitive capacity.

This first year of life is essential for establishing broad connections in the developing brain that are key for cognition, behavior, and communication. Some of these connections are pruned away later in the developmental process, but the richness of connections made in this first year of life is critical. Because chimp, Neanderthal, and *sapiens* brains all appear to develop similarly after the first year of life, this mode of development after year one most likely existed in the common ancestor of all three species. It also means that something novel or derived occurred in our lineage, and that novelty is the globular brain stage in the first year of brain development. It appears that Neanderthals may not have had the neural wiring that *sapiens* did for the kind of cognition and communication we evolved, suggesting that Neanderthals were more than likely quite different in how they viewed the world and communicated about it. These studies also suggest that not only does there need to be an anatomical change in the structure of our voice box for language to arise but there also needs to be a correlated change in brain wiring that increases or at the very least changes cognitive capacity in our species. This change in brain wiring includes better integration of our sense of sound and how we process it when we hear language.

Researchers have sequenced the genomes of several extinct Neanderthal individuals and the close relative known as Denisovan, and they have used the genome data to determine if Neanderthals had the genetic components for language, much in the same way color vision has been examined in Neanderthals (see Chapter 9). There are two major problems with this genomic approach, the first being that although there is most likely a genetic basis for language in *sapiens,* it is very likely an incredibly complex genetic phenomenon. The second problem is that at this time we simply do not have a hold on the actual genetic loci that might be involved except in a couple of interesting but all too general genetic phenomena. The phenotypic complexity of a trait like language is a major roadblock to understanding the genetic basis of the trait. We have already seen this problem manifest itself in this book with synesthetic traits and schizophrenia. Traits like these are difficult to define phenotypically, even with clever tests, and so without a well-defined idea of what the trait really is, genetic dissection is very difficult. In addition, even if one could pin down the phenotype, it is still possible that the trait could be caused by hundreds of genes, all with small effects. Researchers then turn to unique cases in the general population where the phenotype occurs and focus on families where strange phenotypes exist. One "language" gene, FoxP2, has been studied in this way, having been found in a family of people lacking language comprehension. The genetic basis of this trait has been studied at length, and the gene is known to be present in Neanderthals from genome scans. Detailed analyses of the FoxP2 gene sequence indicate that the Neanderthal version of the gene and the *sapiens* gene are identical at the DNA sequence level. This evidence, however, is circumstantial at best with respect to determining whether Neanderthals spoke.

The dissection of language as a phenomenon and its subsequent evolutionary analysis have been the focus of work by Johan Bolhuis and his colleagues. They argue that language itself is not all there is to communication. By pointing out that language, although a form of communication, cannot be equated to it, they come to a more precise definition of what language is and hence can look at the phenomenology of it in a more exact and productive way. Language to them is the capacity to merge ideas and words.

Merging, in its simplest form, is the capacity to recognize objects and actions as coming from self or from others and to be able to sort that out and, more important, to express it in symbols. The argument is that, without merging, language cannot exist. You can still communicate, but you will not have language. It is clear that among living species, merging is a uniquely *sapiens* characteristic, because chimpanzees do not do it. Bolhuis and colleagues also argue that the capacity to merge arose in *Homo sapiens* between seventy thousand and one hundred thousand years ago. Although some have challenged their argument for this timing as being too short and illogically deduced, the timing actually makes sense because it coincides with the movement of our species across the globe and our emergence as the dominant force on this planet.

Two aspects of language to make the discussion relatively complete are reading and writing. Literacy is a very modern development in our species. More than likely writing and reading arose in the last ten thousand years. This short period of time for literacy in our species has had a huge impact on the neurobiology of our senses. The portal for the neural information that is needed for literacy is usually through the retina and hence the eyes. Of course, reading can also be accomplished by the blind using braille, but the tactile neural pathways this information takes in braille is quite different from the pathway for visual reading. Studying the emergence of reading takes a comparative approach. It is not hard to follow the acquisition of literacy in humans by following how children acquire it. The assay of choice is usually fMRI, and longitudinal assays of the brain in children acquiring literacy can be compared across different ages.

As with any comparative approach, the research must be done carefully because some of the inferences can be confounded if age and degree of schooling are not considered. Children of specific ages can't be compared directly because of potential differences in their schooling, so age is not a good starting point for setting up the comparisons. The differences in schooling create a subtle apples and oranges problem for following the acquisition of literacy. Comparing adults who have acquired literacy later in life with those who have not seems to be a better way to get at the impact of reading on the brain. With these caveats about confounding results, some very interesting inferences have been made about the acquisition of literacy.

As with all of the senses, when the initial information enters the brain from the sensory collection organ (in the case of literacy, the organ is the retina) there is an initial rapid processing of the information (fig. 18.3). With literacy there are subtle differences between figures in writing and reading, so the brain suppresses the capacity to lump and instead becomes quite discriminatory with apparently repeated images. It is clear from comparing fMRIs of literate adults with those of illiterate adults that this suppression is more prominent in people who have acquired literacy. These comparisons have also provided finer resolution of the visual pathways of our species and differences between culturally distinct writing systems. Remember that the information from the retina in early visual processing goes through several areas of the visual cortex, specifically the pathways known as V1, V2, V3, and V4. Western writing uses the V1 and V2 pathways to sort out and recognize the characters used in literacy. By contrast, recognition of characters in Chinese writing

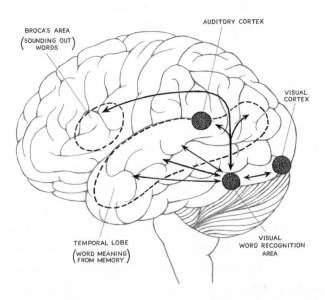

Figure 18.3. The location of specific regions of the brain involved in literacy.

uses the V3 and V4 pathways. The apparent reason for this difference is that Western writing requires knowledge of a rather small number of components in the alphabet. In English, this number is a mere twenty-six bits of information, and this small number of units is best handled by V1 and V2. There are thousands of Chinese characters, though, and these seem to be best processed by the shape-learning V3 and V4 visual pathways. Beyond the early visual processing, the information passes to other parts of the brain.

For our ancestors this processing was needed to be able to recognize things in the outer world relevant to our survival. Our nonhuman primate ancestors did not have writing, of course, so the visual information from writing has a unique and important impact on our species with respect to the ventral visual (the "what") pathway. There is a clearly defined and repeatable region of the ventral processing pathway that shows activity in fMRIs when a literate individual is presented with writing, whether it is Shakespeare or gibberish, in a writing system with which he or she is familiar. This region is the same regardless of language and characters in the language and is called the visual word form area. (Again, this inference is made by comparing people who have acquired literacy to people who haven't.) It appears that this region of the brain becomes highly active with exposure to writing and even to the rudimentary acquisition of literacy.

One of the more interesting developments with the acquisition of literacy in the visual word form area is that this pathway in the brain learns to suppress the tendency to lump mirror images of objects. This tendency to lump is thought to be adaptive, so suppressing it is difficult. That hyena facing to the left is the same as that other one facing to the right, and so there is no need to discriminate between the two. And it follows more quickly that it is wisest to run away from something regardless of mirror image. With the acquisition of literacy there are subtle differences in the characters used. Examples from the Western alphabet include *b* / *d* and *p* / *q*. And hence the adaptive reason for this so-called mirror invariance of our nonliterate ancestors needs to be overcome to acquire literacy. This repositioning of mirror invariance is only one example of the tendency of our species to rewire or repurpose parts of our visual pathways during literacy acquisition to accommodate the unique capability we have to read and write. As noted previously, our species acquired writing and reading very recently. As Stanislas Dehaene and colleagues have elegantly stated, "Literacy acquisition therefore provides a remarkable example of how the brain reorganizes to accommodate a novel cultural skill." It would not be surprising if other regions of our brains have already begun to repurpose neural real estate to accommodate other more modern culturally induced phenomena such as interacting with computers and watching movies and television.

Our species has used language in a myriad of ways. Many uses have had direct bearing on the survival of our species. Indeed, some paleoanthropologists look to language as the spark that sets our species so far apart from all other species on the planet. While writing this book I experienced two relatively improbable events: the Chicago Cubs won the World Series, and Bob Dylan was awarded the Nobel Prize for Literature. One of these events brings me great joy, and the other is puzzling but at the same time intriguing. Being the son, grandson, brother, and uncle of Chicago Cubs fans, and one myself, my capacity for rooting for the underdog is well honed. So, when Bob Dylan, certainly an underdog in the literature world for winning a Nobel, was announced as the Laureate in Literature, followed almost a month later by the Cubs' magnificent victory, it was hog heaven for underdogs. But many in the field of literature questioned the awarding of a Nobel Prize to Bob Dylan as a stretch, and as a result some have even started to question what poetry and literature really are. At the very least, some have started to ask questions about whether Dylan's work really is literature.

Why does poetry affect us differently than randomly chosen words or even specifically chosen words not couched as poetry? The answer to these questions resides in understanding the neurobiology of our senses and how language can incite memories and emotions, much in the same way that music and art affect our emotions and memories. Jonah Lehrer, a former journalist, wryly suggests that the great author Marcel Proust was a neuroscientist. His argument stems from the effectiveness of Proust's writing at inciting the emotions and remembrances of the past, most notably with his madeleine, discussed earlier in this book (Chapter 6). In many ways, if we communicate well, using language, art, or music, we are also neuroscientists. We all have our madeleines, and more important, we all can use language to describe the brain when using language or viewing art or listening to music.

# 19 FACING THE MUSIC

*The Neurobiology of Music and Art*

*"Music is the movement of sound to reach the soul for the educa-tion of its virtue."* —Plato

At the risk of violating the famous comedian, actor, and musician Steve Martin's edict about music that "talking about music is like danc-ing about architecture," I will examine the sensory context of music from several perspectives. What really is music, and does it deviate from language? How did it arise in our species? After all, processes akin to music exist in other species, such as in birdsong or cetacean vocalization. Addressing this question includes looking at the neural and genetic basis of music in our species to understand its origin and possible impact on the nervous system. Also, what are neurophysiological bases of the effects of music and the sensory components involved in our perception of it?

Part of the draw to music as a sensory phenomenon is its general appeal to the brain. Music has been called everything from the thing that soothes a savage breast (Shakespeare) to cheesecake for the mind (Steven Pinker) to anything that is too stupid to be spoken (Voltaire). So, the complex relation of music with our brain and the nervous system is a critical story to understanding how our senses are involved in the cre-ation of perception and how the unique perceptions generated by music affect our inner world of emotion and memory. The major sensory input system that receives music is the auditory system. But some humans can read music and have the capacity to "hear" it simply from reading it. Sadly, my predilection for music is as a casual listener, although I did learn to read music when I was young. Unfortunately, my capacity to read music only goes as far as knowing what the notes are supposed to mean. I can hear music in my mind when it is played in front of me, but reading musical notes now has no impact on what I can "hear." I often have earworms, those annoying tunes that play over and over in the brain. My earworm while writing this chapter was Randy Newman's "You've Got a Friend in Me" from the *Toy Story* movies that my three-year-old son and I love to watch together. The funny thing is that it is just the line from the title that plays over and over in my earworm. Six simple notes, associated with six syllables, have dominated the down-time of my brain for the past week. Eventually, I know I will lose this earworm, but in some ways I hope it doesn't go away. Most earworms are irritating at the least, and distracting and distressing at worst. They are difficult to extract from the brain, and this is why Oliver Sacks calls them "brainworms" instead. My attraction to this one earworm is emotional because it is a bond to all three of my children who have watched the *Toy Story* movies with me, and it has become a part of my memory because even though watching any movie ninety times can be a wearisome task, these particular movies and the earworm itself have become like auditory madeleines to me.

The difficulties that geneticists have encountered in defining music associated with their attempts to unravel the genetic basis of music should be a good example of how hard it is to define music and musical ability. When using genetics to unravel a complex trait, a strong definition of the trait is needed to make any progress. So, what geneticists have done is attempt to dissect the components of music in order to uncover the genetics of this complex trait. To understand the components of music, recall that the music we hear is sound (Chapters 5 and 7), and sounds are essentially waves that cause vibrations. Our ears are inundated with sound, so what is it about music that makes us capable of hearing it and realizing it is music? There is no single characteristic that makes music music. Rather, it is a combination of characteristics about the vibrations being made that our ears detect. It turns out, depending on who you are talking to, that there are anywhere from five to eight major elements of music, and some of them are essential for an understanding of the genetic basis of musical traits. One element, pitch, is usually on everyone's list. We have already discussed pitch as a major element of sound that is based on the frequency with which the sounds are vibrating and the size of the object that is vibrating. So, for instance, the faster the sound is vibrating and the smaller the item that is vibrating, the higher the pitch will be. Pitches exist on a spectrum from low pitch to high and are measured in hertz. Pitch and vibration frequency should not be confused, which means that pitch is sort of subjective. The range of hearing for a healthy young person is 20 to 20,000 hertz, but not all sounds in this range are considered musical. For instance, the range of pitch on a musical instrument like a piano is from $4.37$ hertz ($B_8$ note) to $2,109.89$ hertz ($C_0$ note). People with absolute, or perfect, pitch are able to hear a sound and to identify it as having a specific pitch without a frame of reference. Such people are easy to identify, and the majority of people with absolute pitch who have been studied are musicians.

The brains of people with perfect pitch have also been examined using brain imaging approaches, and comparisons of musicians with and without absolute pitch have been made. It turns out that a specific part of the brain, called the planum temporale, is consistently associated with perfect pitch. Because we have this brain region on both sides of our brain, the first relevant question is do the two sides of the brain differ? Imaging studies clearly show an asymmetry of the left planum temporale and the right planum temporale in people with perfect pitch. The next question then becomes how does the asymmetry arise? The fMRI studies suggest that in people with perfect pitch, there is a pruning of neural connections in the right planum temporale in childhood that renders it smaller than the left. Because the pruning occurs in early years of childhood, and hence before most musical training, it cannot be attributed to exposure to music. Rather, it is more than likely a developmental phenomenon under genetic control. Let's take a look at how imaging the brain on music works and what the imaging tells us.

Cortical thickness is used as a metric in how brain morphology changes. The cortical thickness is used because it is correlated with more white matter and hence more neural connectivity. Greater cortical thickness of the brains of people with perfect pitch is evident in these studies, but the exact nature of the differences has only recently been made clear. The bottom line with respect to neural structure and perfect pitch is that there is an increased neural connectivity in parts of the brain in people with absolute pitch. Specifically, the leftward planum temporale region of the brain appears to be asymmetrically larger in people with absolute pitch. The experiments pinning down this neuroanatomical correlate are done very carefully using only right-handed musicians (remember that handedness sometimes influences the side of the brain where certain neural functions reside). But does this asymmetry exist because the left planum temporale has grown larger or because the right planum temporale is smaller in people with absolute pitch?

The genetic basis of absolute pitch (box 19.1) has been examined using several approaches. The tried-and-true twin study approach showed that there is a heritability of about 0.81 for perfect pitch, which means that there is a strong genetic component to the trait (remember that heritabilities range from 0.0 to 1.0, with values closer to 1.0 indicating full genetic control over the trait). Genomic techniques have attempted to localize the genetic elements involved in absolute pitch. These studies have found several regions in the genome that are linked to absolute pitch, of which one, located on chromosome 8, appears to crop up in most studies that attempt to link absolute pitch to genes. The gene focused on is important in memory and cognition. Other loci have also been forwarded as candidates, and these usually are involved in development of the inner ear.

While not exactly the genetic opposite of perfect pitch, congenital amusia involves the lack of capacity to detect pitch and a lack of ability to remember tunes. As its name implies, it runs in families, and its neuroanatomy has been worked on. Specifically, amusia has a major impact on the auditory cortex and in this context has a very different neurobiological etiology and indeed does not involve the same genes that allow absolute pitch. To date, a genetic basis for amusia has been implied, but the genetic locus affected is not known.

Another method that can be used to discover genes that might be involved in musical aptitude or preference is to ask where in the genome natural selection may have had an impact in musicians. This approach looks for genes in the genome where DNA sequences have changed in a specific extraordinary way that implies natural selection. In some tests

The specific location of the linked gene is 8q24.21. Genomic locations start with the number of the chromosome, and they next list whether the locus is on one end of the chromosome (p) or the other (q). Human chromosomes in general have a centromere in them that separates the short end (p) from the long end (q). Finally, the location of the locus is given by coordinates much like a ruler. This gene at 8q24.21 is thought to be ADCY8, or adenylase cyclase 8, which has a very specific cellular function in the cell membrane. It is also a gene that is thought to be involved in memory and cognition. Loci involved in absolute pitch might also be involved in the development of the inner ear, neural connectivity, and development. If loci involved in neural connectivity are involved, this might support the contention that absolute pitch is all about the pruning of connections in the brain.

for selection, the genome is scanned region by region to see if DNA sequences occur in nonrandom distributions. If a gene does have a profile of change where it statistically appears to be unusual, it is tagged as a potential gene involved in a specific trait. One such study looked at about 150 Finnish individuals, using musical aptitude as a means to sort through unusual signals. Their results suggest that several genes show signatures of natural selection, and some of these are involved both in the development of the inner ear and in aspects of cognition. Interestingly several of the genes that were discovered have unknown functions in humans, but the same genes are responsible for perception of songs and song production in songbirds. As I have pointed out before, these genetic studies are only as good as the characterization of the phenotype is. If one has hard time determining the phenotype, then finding the genetic correlate is going to be either difficult or misleading. Absolute pitch and amusia can be detected with simple tests, so the reliability of the genetic bases of these traits is pretty strong. Other traits that researchers have explored with regard to music are harder to pin down as phenotypes. Traits like musical aptitude, ability, and preference are a bit harder to examine, but looking at these blurrier traits gets us closer to why and how music soothes a savage breast.

To quantify these fuzzier musical abilities, researchers have turned to standardized tests that allow them to place people on a spectrum of values, such as the Karma Music Test. In this test, people are exposed to short, abstract sound patterns that form hierarchical musical structures because they are repeated. The person will hear several different hierarchical patterns and is then asked to discriminate among them. How accurately someone can describe the differences in the hierarchical patterns can be placed on a scale. The test is interesting because it was devised to weed out musical training as a factor in the scores. Other tests that are applied are called Seashore tests and the pitch production accuracy test. The Seashore tests consist of the subject hearing pairs of sounds with slight differences in pitch and timing. The person is then asked to discriminate between the pairs of sounds. The tests detect simpler aspects of the sensory capacity for musical aptitude such as pitch and timing. The pitch production accuracy test starts with the person hearing a sound with a specific pitch through headphones. The subject is then asked to sing the note replicating the pitch from the headphone. The performance of the person is then easily graded to give a quantitative measure of musical ability.

Musical preference is a different story when it comes to the tests for quantifying it, because the subject's opinion of a particular kind (a genre) of music is quantified. If the participant is an infant, the attention trick described in Chapter 7 for synesthesia in infants is used. The Short Test of Music Preferences is the most widely used approach. This test presents fourteen to twenty-three music genres to the listener, who is asked to rank the genres on a scale of 1 to 7 (1 = strongly dislike and 7 = strongly like). Some researchers have gone as far as asking subjects to write essays describing musical genres or even songs. In one study, researchers pored over more than 2,500 essays written about music genres to assess the musical preferences of the study participants.

With the usual caveats about how this kind of work is done, some interesting ideas about musical aptitude and preference have been made. For instance, using the quantification of musical ability, several potential traits associated with musical aptitude have been looked at genetically. These subtraits include recognition of pitch and rhythm, music memory, music listening, singing, and musical creativity. For recognition of pitch, several loci have been pinpointed by genomic studies. Many of the genes discovered in this way are involved in the development of the nervous system or maintenance of neural tissues. A couple of the genes are specifically important in the development of the inner ear for structures like the cochlea and for the proper development of the tiny hairs in the inner ear that detect the external vibrations of sound. One gene in particular codes for a protein that appears to be important in pitch and rhythm, musical memory, and music listening. The gene's name is AVPR1A, and it is technically known as a vasopressor, a protein that acts in regulating the amount of water the body holds and with blood pressure. Vasopressin is another name for AVP, a hormone that has been suggested to be associated with autism and is also thought to affect interpersonal behavior. Another gene thought to be involved in musical aptitude is called protocadherin, which is a membrane protein that regulates cell-to-cell adhesion. It is important in the structure of the cochlea.

Musical preference has also been studied in the context of how music affects our emotional makeup. For instance, there is a lot of work on how musical preference intersects with personality and, by extension, whether musical preference can be a predictor of personality. Some studies show a clear correlation of personality with musical preference. A large study used the five-factor model of personality that attempts to place personalities on a scale with the following descriptors: extraversion, agreeableness, conscientiousness, neuroticism, and openness to experience. These personality descriptors were examined in the context of music preference for several genres of music. The result of the survey is that lyrics and genre are correlated to personality, with personalities like openness correlated to the wilder music like punk and death metal and extraversion correlated with pop music. Indeed, researchers suggest that the one (personality) can be used to predict the other (genre) and vice versa.

Other studies reveal similar patterns indicating that individuals who are open-minded to experiences in their lives are more prone to have preferences for complex music like classical music and music considered on the cutting edge like punk rock. On the other hand, extraverts will go for pop music like the boy band music and rhythmic music like hip hop. Delinquency has also been studied as a part of the story. Using the reporting systems described above and a database of the delinquency of participants, Tom ter Bogt, Loes Keijsers, and Wim Meeus have shown that "loud, rebellious and deviant music," such as heavy metal, goth, and punk, as well as rap and techno house music, are correlated with minor delinquency in older adolescents. Classical music and pop show no link to delinquency. These data together indicate that early musical preference might point to the direction of minor delinquency later in life. I hate to admit it, but perhaps my parents were right when they banned my playing of "The Pusher" by Steppenwolf in our house. The rock and protopunk music I listened to as a kid never landed me in jail, but I am sure that it did lead to some delinquent behavior on my part. But is delinquent behavior all bad? If you have ever been the parent of a rambunctious sixteen-year-old, then the answer might seem obvious. But some researchers suggest that the real link of music like punk and heavy metal is to openness and propensity to explore.

The kinds of genes that have been pinpointed so far that are involved in musical aptitude show us a lot about how music works in the brain. The psychological testing done to understand music preference also tells us something deeper about music. Music, in general, enters the brain through the ears, so genes that are important in the development and structure of the inner ear are involved in the primary processing of sounds like music. Once in the brain, music is processed into its component parts, and from these the brain then dictates how specific kinds of music will affect us, by inducing emotions and memories.

Thomas Schäfer and colleagues conducted a survey on more than eight hundred subjects and were able to suggest that people listen to music to "regulate arousal and mood," to "achieve self awareness," and as an "expression of social relatedness" (box 19.2). The first two functions were more influential than the third. It should be noted that the study focused on German-speaking people over a large range of ages (eight to eighty). Although the results of this study are very interesting with respect to defining how music is used by people, music preference itself might be highly cultural, and so it would be interesting to see how other cultures view music in this functional context.

To address the issue of cultural input into musical preference and music creation, Patrick Savage and his colleagues examined music in nine geographic regions of the world. Thirty-two musical features were examined in 304 music recordings from across the globe. Taking the cultural context out of music results in the discovery of no clear diagnosable universals across the different areas of the globe, and this suggests that

## BOX 19.2 | WHY WE LISTEN TO MUSIC

Thomas Schäfer suggests that music listening and preference have more than a hundred functions. By asking study participants a series of 129 questions such as the following, Schäfer and his colleagues were able to dissect preference for music into an equal number of functions.

I like music (please score on a scale of 1 = I do not at all agree to 6 = I agree completely)

\_\_\_ Because it gives me intellectual stimulation.

\_\_\_ Because it gives me something that is mine alone.

\_\_\_ Because it gives me goose bumps.

\_\_\_ Because it addresses my sense of aesthetics.

\_\_\_ Because it reminds me of a particular person.

\_\_\_ Because it makes me feel my body.

\_\_\_ Because I can enjoy it as art.

. . . and so on, for a total of 129 questions relevant to 129 possible functions of music.

---

music may not be the universal language of humankind that many think it is. However, eighteen of the thirty-two features of music do show a statistical global correlation, with ten of these being found in a network of relatedness, which simply means that they are connected to each other more than they are to other features. The statistical features are based on attributes of music that have been suspected to be universal, such as pitch and rhythm. Other features discovered using this approach are not commonly thought to be universals of musical preference, such as performance style and social context. Cross-cultural contexts become very important when attempting to define music and when correlating music to function, however.

The physiological response of the body to music through its action on the brain has also been studied. The experiments to examine this aspect of music involve contrasting the physiological response of people to stress after listening to three background sounds—Gregorio Allegri's "Miserere" (a choral piece written in the seventeenth century for the Sistine Chapel), the sound of rippling water, and silence. These sounds are followed by experimenter-induced stress to the subject. Each participant is then measured for physiological markers of stress, such as cortisol levels, heart rate, and sinus arrhythmia, as well as self-reported stress and anxiety levels. The first part of the survey sounds okay and perhaps even fun, especially if you get to listen to the "Miserere," which is a beautifully soothing piece of music. But the stress part sounds a little like torture. The researchers doing this survey thought of two of the most stress-inducing situations an adult could go through without causing psychological havoc—a job interview and the task of solving a difficult arithmetic problem in front of an audience. I can attest to the stress induced by the second task of doing math in front of people. When I started my academic career, I took a position as an assistant professor at Yale University. I had never taught before but felt some confidence in my ability to explain population genetics, which can be very mathematical at times. During my first lecture at Yale, I decided to derive a mathematical equation without notes for an undergraduate genetics class. I started okay with the basic parts of the equation but quickly got lost, and this in front of a hundred Yale undergrads. My stress level rose massively as I tried to recover the missing parts of the equation, and then it happened. I was booed by the students, and my stress level shot through the roof. So, while it might be slightly tortuous to induce stress in this way, I can attest to it working quite well at inducing stress. Of course, the idea of the study is to see if the soothing "Miserere" music and the potentially soothing sound of water could alleviate some of the stress applied after listening to the two sounds. The results indicate that listening to "Miserere" (relaxing music) before the application of stress does not make one immune to stress. However, the musical treatment before stress induction means that an individual will recover faster from the induced stress.

Stress is only one emotional and physiological response that can be modulated by music. Using a clever animation to reflect emotions, Thalia Wheatley and her colleagues examined the emotional context of music in Dartmouth College students and members of the Kreung tribe in Cambodia. They used a clever device called Mr. Ball to gauge emotional level and musical preference.

A red bouncing ball, Mr. Ball can be animated and his physical appearance manipulated. His physical appearance is controlled by five sliding bars, which a person can move to represent an emotion. In one animation, a group of subjects was asked to manipulate Mr. Ball using one of the five sliders to show happiness, sadness, peacefulness, anger, or fear. In a different animation, the sliders represented aspects of music matched with aspects of motion. So, for instance, slider number 1 controlled the rate of bouncing and the speed of the music. Slider 2 controlled jittery movements and the jitteriness of the music. Smooth movement and consonant musical sounds were controlled by the third slider. The fourth slider controlled size of move of steps and move of notes, and the fifth slider controlled the direction of the steps, or whether the music moved to higher notes or lower notes. The key to the experiment is that one group of subjects would use Mr. Ball's movement (and not hear music) as a way to express the emotion, and the other would use music (and not see movement) to express the requested emotion.

Interestingly, when a person was asked to make say, an angry Mr. Ball, the sliding bars were placed in nearly identical positions for music and for the movement of Mr. Ball (fig. 19.1). This result was also obtained when Cambodian people from the Kreung tribe were asked to do the same experiment, indicating a cross-cultural context for the task. This experiment actually addresses questions about the cross-modality of sight and sound, but in an emotional context. The result suggests that both motion and music activate brain regions connected to emotions that are deep in the brain in the limbic system, where emotions are processed. This undeniable link of emotions to music is also demonstrable with literature and art. The differences are in how the sensory information gets into the brain and where it travels from this entrance.

In 2005, two years before Jonah Lehrer suggested that Proust was a neuroscientist, Patrick Cavanagh, a neurobiologist, pointed out in the

Figure 19.1. Mr. Ball in different guises (happy on the left, angry on the right), as imposed by subjects of Wheatley and colleagues' surveys.

journal *Nature* that artists have had a centuries-long and wonderful hold on neurobiology and on the visual process. He presented several intriguing examples from Renaissance art where artists tricked the senses with shading or lighting. The examples involve improbable lighting or shading techniques that the observer rarely if ever notices. These tricks include using shadows and shading of buildings to create perspective, at the cost of impossible physical attributes of the shadows and shading. He made the very interesting point that certain attributes of art have not changed for thousands of years. For instance, line drawing was established early on in the history of known art. The amazing renderings of animals in cave drawings like those from Lascaux, France, from fifteen thousand years ago are similar to drawings of animals from the fifth century and also to modern line drawings. Cavanagh points out that early artists experimented with drawing lines that their viewers would be able to perceive and identify as the objects they were drawing, and they did this in the earliest known portraits of animals. This technique is also evident in sculptures and goes back just as far as the Lascaux drawings, as evidenced by the superb carving of a horse from the rock shelter known as the Abri de Cap Blanc in France (also fifteen thousand years old).

Once this technique was learned, the rest becomes what art is all about—creating interesting and intriguing and baffling ways of playing with it. As V. S. Ramachandran, the well-known neuroscientist (see Chapter 12) and contributor to ideas about art and neuroscience, says, "The purpose of art is to enhance, transcend, or indeed even to distort reality." This basic knowledge is an artistic technique that cannot be "unlearned" and is emblazoned in the brains of artists and their patrons. Ramachandran has been criticized for the reductionism in his explanation of art as a neurobiological process because he minimizes emotion, memory, and intellectual intention. But what Ramachandran suggests is a good start, because after all, art begins with our senses. If there are complicated feedback loops with emotions and memory and intention, then this is all secondary to the initial impression art makes on us. Cavanagh puts it into perspective with the following statement: "Discrepancies between the real world and the world depicted by artists reveal as much about the brain within us as the artist reveals about the world around us." And so, studying artists and their patrons while viewing art has become an interesting and productive way to discover not only what art is made of from a neurobiological perspective but also how our brains work in general.

Other artistic techniques, such as transparency in paintings, the use of two dimensions to convey three-dimensional scenes, incompletely drawn out art that begs the viewer to fill things in, and reflections, also have a neurobiological tinge to them. Here again Cavanagh gives some beautiful examples of art with neurobiological explanations for the visual effects they produce. (Readers should consult the original article in *Nature* to get a sense for the neurobiologist-artist connection.)

One of my favorite art movements is cubism. In many cases, cubist art shows just enough of the subject to get the brain to identify what might be the information the eyes have collected. I enjoy cubist works because they induce a basic biological response in me when I view them and then let my imagination go wild. This capacity to take cubist artwork and identify the objects in it is a neurological function that our species more than likely evolved in common ancestors far in the past. Evolutionarily it is important for any visual information the retina collects to be identified rapidly so that we can decide whether to run away from the object, eat it, or try to mate with it. Sometimes in nature retinas only collect information for fragments of the object: think of the proverbial snake in the grass or a hyena snout sticking out from behind a rock. But organisms still need to make quick decisions about the object that might be essential for survival. Cubism has exploited this basic visual neural function of "filling in" and has manipulated this function to produce some of the most intriguing and inspiring artwork around. Finally, cubism doesn't appeal only to our so-called lizard or reptilian brain. It has a broader and more widespread effect because waves of signals are sent throughout the brain after the object is recognized in the inner part of the brain. More than likely this information travels to the same places in the brain that information on an abstract piece of art, such as an action painting by Jackson Pollock, or a Renaissance oil painting, such as Leonardo da Vinci's *Mona Lisa,* does for postprocessing.

One way to dissect what is involved in art perception is to simulate art by using obvious aspects of art that are well-defined to produce art and then dissect the response that people have to the simulated art. Using computer-simulated art generated by what is called the Painting Fool, researchers have made some inroads into dissecting how art is made at the neurological level.

This computer program is the brainchild of Simon Colton and has its own website for producing simulated art. After using it for a while, I got the sense that the Painting Fool has a personality and certain degree of pride in its work, and it makes some very interesting visual products. And indeed, the Painting Fool has been programmed to simulate the rational moves and techniques of an artist as a piece of art is generated. Researchers can tip this approach on its head and program the Painting Fool to be irrational and ask the question, What happens? Oddly enough, the Painting Fool produces very trippy, hallucinogenic art when the rational rules of art that it usually works by are tweaked. This result suggests that creativity and hallucinations are connected in an interesting and compelling way in art. One need only look to the trippy art of the surrealists to see the logic behind this approach.

Artistic capacity has been observed to run in families, suggesting a genetic component. Any genetic study focused on art almost obligatorily concerns visual input and the assessment of aesthetic preference. The retina usually is the portal through which information about art travels into the nervous system. Hence, any trait tightly linked to enhanced functioning of the retina might be similar to perfect pitch with music and hence perhaps a direct connection to art. Because we know a lot about the structure of the retina, looking at the component parts of this structure in our eyes could be a good inroad to linking genes with art perception and preference. The first and foremost requirement for the structure of the retina in viewing art is the proper development of rods and cones. Without these cells that collect light waves, blindness results. This doesn't mean that blind people cannot enjoy art, because they still have the sense of touch. A visit to see Leonardo da Vinci's *Last Supper* fresco at the convent of Santa Maria delle Grazie in Milan is a stunning visual experience. But off to the side in the grand room where the *Last Supper* is painted lies a three-dimensional relief sculpture of the famous scene. It is placed there specifically for the blind, and even for someone with sight, if one closes the eyes and feels the relief with the hands the experience is also very moving.

The next structural aspect of the retina would be what kinds of rods and cones exist in the retina, and this aspect of the retina concerns the opsins embedded in the rod and cone membranes. As we saw in Chapter 9, some people, known as tetrachromats, have extra opsin genes that make proteins that are sensitive to colors that normal people, or trichromats, cannot see. In addition to perceiving color, the opsins in rod and cone cells collect information on shading and depth perception. Some researchers have examined people who are both tetrachromats and artists to see if there is a correlation of this trait with art production and perception. These studies suggest that tetrachromats experience and produce art in different ways than normal trichromats. This doesn't mean that tetrachromats make better art or are better at perception than us normal trichromats; rather, the paintings they produce are done in unique sensory manners. Other aspects of vision also enhance how we collect and interpret light waves. The visual pathway in the human brain is very well known, and perhaps some of the neural connections in these areas of the brain will eventually be shown to have an impact on how we make and perceive art.

Our senses take information from the outside world and transform it to perception and then to meaning. It should be clear from the examples in this chapter that the initial processes used to get the information into the brain are very similar in most higher organisms, especially vertebrates. However, what should be equally clear is that our species turns the initial information from the outer world into something unique in comparison to other organisms on this planet. It is stunning, though, to realize that our species is continually attempting to improve our senses and therefore to improve our perception of the outer world. We basically have no limits to how we can and will perceive the outer world. The lack of limits means that we have the power to help correct deficiencies in some of our fellow humans with respect to the senses. But it also means that various parts of the universe that we have not so far been able to see, feel, taste, smell, and hear will eventually be made perceivable to us by development of new technologies that enhance those senses.

## 20 NO LIMITS

*The Limits to What We Can Sense and the Future of Our Senses*

He appears on the TED stage, and without knowing who he is, my first reaction to his attire is, "What was he thinking when he left the house?" His shirt is bright blue, his jacket is pink, his pants are brilliant yellow, and on his feet are black-and-white saddle shoes. I can only guess what color his socks are. His name is Neil Harbisson, an artist and one of the most famous monochromats on the planet. Remember from Chapter 9 that a monochromat can see the world only in shades of black and white. So Harbisson sees the world, as he likes to point out, as if he were watching TV in the 1950s—that is, in tones of black and white and shades of gray. What is remarkable about Harbisson is that he calls himself a cyborg. He wears a cameralike device on his forehead that is connected to the back of his brain and makes sounds when it is pointed at something that has color. A high-pitched squeak is emitted when the camera focuses on a dirty yellow sock, and a lower, mellower sound emerges when Harbisson holds a red handkerchief in front of the camera. He has learned to use the sound vibrations from the device to observe the colors of objects. The cyborg in him has brought color to his otherwise black-and-white visual life.

As we have seen throughout this book, our senses do have physical limits, placed on them by the structures that collect and interpret information from the outer world. But our species has not been limited by these constraints on our senses. Our capacity to see is limited because the human retina can collect light in only a narrow range of electromagnetic wavelength. The wavelengths that exist in nature range from 100,000 kilometers (100,000,000,000 m) to 1.0 picometers (0.000000000001 m), or more than twenty-four orders of magnitude. Our visible range is only over a few hundred nanometers. In other words, humans' visible biological range is only a paltry 0.00000000000000000000001th of the entire spectrum. Yet we know that light at these other wavelengths exists, and we have even developed aids that help us visualize the light or the product of light at these other far-ranging wavelengths.

For sound, we can hear in the range of 20 to 20,000 hertz. That's more than three orders of magnitude. All other sounds outside this range are undetectable to most humans. Yet again we know these other sounds exist, and we have made instruments to detect them even though our own physical neurological machinery can't. As we saw with odors, we have a relatively large number of smell receptor genes that translate into a capacity to smell over a fairly large range of odors. One estimate mentioned in this book is more than $10^{12}$ (more than a trillion). This is a huge number of odors, but nonetheless only a very small range of the total number of kinds of odors that exist on our planet. Yet once again we can characterize these odors that are odorless to our neurobiology. Taste and olfaction are similar in their neurobiological mechanism in that both are chemosensory, yet taste has far fewer receptors and only five real categories of what can be detected by the gustatory system. Yet once again we know that the range of molecules that interact with our taste buds that are out there is much greater than what we taste. We simply don't have receptors for these other molecules, but we know they exist and we develop methods whereby we can characterize them. With fertile minds such as the one Charles Spence possesses (see Chapter 15), it is possible that we will find a way to characterize the better than trillion odors that we might be able to sense.

Part of the limited range of the senses we have is the result of how well tuned our biology is to our senses. But what is of equal importance to our modern human existence is that when our scientific, cultural, or social needs have exceeded the range of our biological senses, we have figured out a way not to be limited by our biology. X-rays, sonar, magnetic resonance imaging, and microscopy are only a few of the hundreds of innovations humans have made that allow us to expand beyond evolved biology every day.

My favorite example is DNA sequencing. Up until about sixty years ago, our species hadn't even identified DNA as the hereditary material and had no idea what constituted it or how it was configured. Up to the 1950s, researchers in chemistry and physics were discovering spectacular information about the structure of "invisible" compounds such as proteins and carbohydrates (the constituents in living organisms), which was another equally stunning feat extending our visual capacity. But this knowledge of chemistry was extended because humans were reaching outside of the range of their senses. X-rays were being used to visualize the three-dimensional structure of molecules, and indeed one of the big steps to deciphering the physical structure of this important molecule was the use of X-rays to examine crystal structures of molecules like DNA. In 1953, James Watson, Francis Crick, Maurice Wilkins, and Rosalind Franklin were able to pin down the structure of DNA as a double helix with a diameter of 10 angstroms. This diameter is several orders of magnitude outside the range that we can normally see with our eyes. But the proposed structure was important because, as Watson and Crick so wryly pointed out in their 1953 *Nature* paper, it did not escape their notice "that the specific pairing we have postulated immediately suggests a possible copying mechanism for the genetic material." Seeing the double helix and making the leap to how it works as the genetic material required that our species learn to see things in the X-ray range of electromagnetic radiation. With this information in hand, the next steps were to figure out how DNA did the trick for heredity. I have shown you in several parts of this book what DNA sequences look like and why they are important. But how really do scientists "see" those nucleotides (the Gs, As, Ts, and Cs) that make up the sequences? The size of a nucleotide is way too small (even smaller than the 10-angstrom diameter of the DNA double helix) to be able to see, even through the most powerful electron microscopes scientists use. Reading the recipe of life, as our genomes have been called, is a great example of overcoming sensory noise and our limited range of sensing to interpret a set of devised sensory cues that would make no sense whatsoever to any other species on the planet. For that matter, the way we see DNA is so esoteric that only maybe a few million people on the face of the Earth really can take the sensory information that has been corralled to read the sequence of the genome and make sense out of it. Basically, scientists have used the chemical nature of nucleotides and DNA to amplify its signals to give us a visual string of nucleotides that we interpret as the DNA sequence. Instead of light waves producing the sensory input, chemical reactions are used. These are then interpreted into visual output on a computer screen that our eyes can then see. It sounds like magic, but it really involves some basic but ingenious inventions to read these small molecules way out of our visual range. It's not just the ability to see small objects that we have developed. Most of modern astronomy and astrophysics take data that have very little to do with our evolved visual capacity and convert them to images that we can see and interpret. Although an optical telescope enhances the ability of the retina to absorb light waves from items in the night sky, a radio telescope takes radio waves several times the length of those wavelengths the brain uses to interpret light and converts the radio wavelength data into stunningly informative images of planets and suns that are light years away from Earth.

As a graduate student in Saint Louis in the late 1970s, I well remember needing to compute a solution to a data set. I had only about 100 data points but needed to compute the possible solutions for the data set for exactly 10,395 possible permutations. At that time, the solution to the problem would be accomplished with a pencil and pad of paper, much like in the 2017 Oscar-nominated movie *Hidden Figures*. A group of mathematically inclined humans would take the data and, for all of the 10,000 or so permutations, would undertake the calculations. My thesis project didn't involve national security, nor was it relevant to NASA, so I did not have the luxury of an army of dedicated pencil pushers to compute the solution. I turned to one of the newer possible ways of computing the solution—a computer. This was in the early days, when programs and data were punched onto cards and read into a mammoth computer to start the computation. There was then the long wait to get your green-and-white printout for the job. Clunky to say the least. My first graduate student wrote his thesis in the late 1980s on an Apple Macintosh and used this computer to do most of the same kinds of calculations I did for my thesis on the Macintosh. His graduate students used iMacs to do their work, and his grad students' students used iBooks. All of this over a period of about fifteen years. Now our current generation of graduate students use MacBook Pros that are tens of thousands times more powerful than iMacs and iBooks, and these more modern computers can connect to clusters of processors that give them computing power billions of times more powerful than what my first student could use. This example simply shows that computing in science has expanded as a function of time in a pattern known as Moore's law.

Gordon Moore in the 1960s astutely recognized that computing power will double every year. Computing among the consumer population has increased, too, and in essence has become more personalized than most people in the 1960s would ever have dreamed, except for those who pursued the development of the personalized computer, such as Steve Jobs and Bill Gates. This personalization has changed the way we live from day to day, but it has also changed the way we humans sense the outer world. And given that Moore's law appears to be a real phenomenon, we should attempt to anticipate the rise in power of computational approaches and perhaps even anticipate some of the novel changes our senses will be exposed to as a result of the surge in computing.

The average adult in the Western world faces a computer screen or a smartphone screen for about ten hours a day, according to a 2016 Nielsen survey. Given that we sleep about seven to eight hours a day, this means that more than half the waking day in many cultures is spent staring at a computer or smartphone screen viewing virtual images the whole time. We are only beginning to understand the impact of this changed sensory realm on the human condition. In a direct comparison of reading comprehension among tenth graders, researchers in Norway assessed the difference between reading on a computer screen versus old-fashioned hard copy. The surprising result was that these students comprehended the written word on paper much better than on screen. Why this might be so is not well understood, but it does point to a possible dichotomy in the way we learn and comprehend reading as humans. Reading comprehension is a downstream effect of vision, and some researchers are concerned about the long-term impact that computer and smartphone screens might have on the human visual system in a more upstream manner. Humans *did not* evolve to peer endlessly at a small, light-emitting rectangle. In fact, our field of vision is much greater than that small smartphone screen you scan every day for hours. How this restriction of the visual field is affecting our eyes and their potential evolution is a subject that needs attention. Vision is not the only sense that is under assault by modern life. As noted in Chapter 11, modern humans experience sounds, sound levels, and ranges of sounds that our ancestors never had to. How we adapt to this changed auditory world is also a subject ripe for study.

There is one area of modern life enhanced by computers where researchers have spent some time examining—gaming. Young people today spend inordinate amounts of time playing computer games. Fatima Jonsson and Harko Verhagen have taken a multisensory look at the impact of gaming events on participants. Their conclusion is that even though the games themselves are extremely visual and auditory, the whole repertoire of the senses is involved in the gaming experience, all the way down to taste and smell. In fact, the auditory aspect of video gaming doesn't only emanate from the computer but also involves the sounds from around the screen such as other individuals cheering and screaming as a result of play. And it would not be surprising to note that the olfactory and gustatory impact of gaming is strongly influenced by fast foods and soda. These upstream or basic sensory effects are somewhat easy to characterize, but researchers have also tried to examine downstream neurological effects of sensory stimulation via video gaming. Psychologist Angelica Ortiz de Gortari has carved out a niche in this regard and has studied what she calls game transfer phenomena as a result of intense game playing. For some gamers, their sensory experiences are so intense (and their mental states are susceptible enough) that they experience pseudohallucinations as a result of the gaming. They also incur visual aftereffects that can potentially cause them to misperceive the real world around them. It gets worse with more and more prolonged gaming, and it affects not only visual and auditory sensory perception but tactile and perhaps olfactory sensation, too.

Virtual reality (VR) has also become a modern-day reality. The 2016 holiday season doubled the number of VR headsets in British households to 20 percent, and other Western countries are following close behind. Vision and hearing are not the only senses that VR targets. Entrepreneurs and engineers such as Adrian David Cheok have suggested that all five of the Aristotelian senses can be incorporated into VR apparatuses. But what will the effect of VR be on our senses and sensory perception of the world? It turns out that we might be preadapted to a VR world. Andrea Stevenson Won and her colleagues suggest this interesting possibility, due to a phenomenon called homuncular flexibility. Disembodiment caused by VR can be disorienting, causing physiological and psychological effects. But homuncular flexibility can overcome these problems and enhance the VR experience. The flexibility idea is based on early experiments on phantom limbs. People who have lost limbs often feel extreme pain in the area where the lost limb used to exist. Neuroscientist V. S. Ramamchandran, whom we visited earlier in this book in our discussion of Capgras syndrome (Chapter 12) and the neurobiology of art (Chapter 19), asked people with missing limbs to place their injured limb into a box containing a mirror positioned along the middle of the box. He then asked the person to look into the mirror so that the injured limb visually appeared to be replaced by the uninjured one. What happened is that when the person moved the uninjured limb, he or she saw the illusion of two normally moving limbs. After the person experienced this illusion, the phantom limb syndrome was either reduced or eliminated.

Another example of homuncular flexibility is the rubber arm illusion (fig. 20.1). In this limb illusion, the person seated on the left places one limb below the desk where it is not visible. A disembodied rubber arm is placed on the desk. Simultaneously, the person on the right strokes the tips of the fingers of the hidden hand and of the rubber hand. The person on the left will soon establish a sense of ownership of the rubber arm. Sense of ownership means that if a hammer is brandished by the guy on the right threatening to hit the rubber arm, then the person on the left will flinch.

Both the phantom limb and rubber arm illusions demonstrate that people can be led to reconfigure their body image by visually tricking their brains. In other words, our brains are flexible enough to reconfigure our sensory homunculus (discussed in detail in Chapter 3). Virtual reality is like making your entire body a "phantom limb," or like conditioning your avatar to be like a rubber arm. Our brains are fully capable of this.

Virtual reality, smartphones, technology in modern cosmology, DNA sequencing, and other extensions of our perception by modernity also

Figure 20.1 The rubber arm illusion. The person on the left can be tricked into thinking that the rubber limb is his or hers.

have huge impacts on how we view our world and raise questions about how the human mind will deal with this brave new world. The problem of the human mind is a huge one. Thousands of books, millions if not billions of words, and much human thought have gone into trying to understand the human mind. The approach in this book has been to address what neuroscientists call the easy problems of the mind or of consciousness. These easy problems include localizing where our perceptions of the outside world originate and how this perception works. In some ways we are in a bit of a tough place when discussing these topics. We know, as Francis Crick so eloquently wrote, that "a vast assembly of nerve cells and their associated molecules" are responsible for the mind and the emergence of consciousness. And what the easy problems deal with are those physical aspects of perception caused by nerve cells and molecules. But the holy grail is located in the realm of what neuroscientists call the hard problem of consciousness. This problem is indeed hard, because any answer to it wants to link an emergent property of our neurobiology (mind) with physical, molecular, and chemical information. The spot between the rock and a hard place is that we need the easy problems solved to shore up any ideas we have about the hard problem, while the answers to the easy problems don't get us all the way there. Here I have tried to take an evolutionary approach to understanding the senses and the easy problems of the mind.

This approach can be very illuminating. It is straightforward to reconstruct some of the evolutionary history of sensory processing in our species and our close relatives. An example can be seen in a book my colleague Ian Tattersall and I wrote in 2012. Using Antonio Damasio's ideas about emotions, we reconstructed the evolutionary history of emotion in animals. This reconstruction suggests, not surprisingly, that our species is unique in how we deal with the outside world emotionally. Likewise, the evolutionary view of how our senses evolved indicates that we are in many ways unique in how our sensory information is processed. It is also evident from Ian Tattersall's writing on human consciousness that language and all of the very human things we do around language are essential developments in the emergence of the mind in our species. Although we may still have a huge way to travel to unlock the hard problem, much of the easy problem is unraveling before us as a result of modern neurobiology. It is therefore critical that we keep in mind the no-limits aspect of our sensory development and change as a species. It may be the key to understanding the hard problem of the mind.

# LITERATURE AND FURTHER READING

## CHAPTER 1 | THE BRAINLESS MAJORITY

For a wonderful review of the early life on Earth, see Lane (2010). Microbial life on our bodies is described nicely by Dunn (2011) and microbial distribution in general in DeSalle and Perkins (2015). Dunn's provocation is described in Dunn (2009). Galileo's take on the senses is described in the scholarly work of Piccolino and Wade (2008). Anishkin et al. (2014) present their argument for mechano-sensation as the original sense, and Bassler and Miller (2013) describe quorum sensing. Magnetobacteria are discussed in Uebe and Schüler (2016), and the dancing bacteria can be viewed at https://www.youtube.com/watch?v=3uUL4o0M6KI. Horizontal transfer in magnetotactic bacteria is discussed in Lefèvre and Bazylinski (2013). The case for plant neurobiology is made in Baluška et al. (2006), but Pollan (2013) makes the case for plant intelligent behavior.

Anishkin, Andriy, Stephen H. Loukin, Jinfeng Teng, and Ching Kung. 2014. "Feeling the Hidden Mechanical Forces in Lipid Bilayer Is an Original Sense." *Proceedings of the National Academy of Sciences* 111, no. 22: 7898–7905.

Baluška, František, Stefano Mancuso, and Dieter Volkmann, eds. 2006. *Communication in Plants: Neuronal Aspects of Plant Life*. Springer.

Bassler, Bonnie L., and Melissa B. Miller. 2013. "Quorum Sensing." In *The Prokaryotes: Prokaryotic Biology and Symbiotic Associations*, ed. Eugene Rosenberg et al., 495–509. Springer.

DeSalle, Rob, Susan L. Perkins, and Patricia J. Wynne. 2015. *Welcome to the Mi-*

crobiome: Getting to Know the Trillions of Bacteria and Other Microbes In, On, and Around You. Yale University Press.

Dunn, Rob. 2009. *Every Living Thing: Man's Obsessive Quest to Catalog Life, from Nanobacteria to New Monkeys.* HarperCollins.

———. 2011. *The Wild Life of Our Bodies.* HarperCollins.

Lane, Nick. 2010. *Life Ascending: The Ten Great Inventions of Evolution.* Profile Books.

Lefèvre, Christopher T., and Dennis A. Bazylinski. 2013. "Ecology, Diversity, and Evolution of Magnetotactic Bacteria." *Microbiology and Molecular Biology Reviews* 77, no. 3: 497–526.

Piccolino, Marco, and Nicholas J. Wade. 2008. "Galileo's Eye: A New Vision of the Senses in the Work of Galileo Galilei." *Perception* 37, no. 9: 1312–1340.

Pollan, Michael. 2013. "The Intelligent Plant." *New Yorker,* December 23 and 30.

Uebe, René, and Dirk Schüler. 2016. "Magnetosome Biogenesis in Magnetotactic Bacteria." *Nature Reviews Microbiology* 14, no. 10: 621–637.

## CHAPTER 2 | BRAINS AND PRAINS

For more information on placozoa, see Eitel et al. (2013). Information on these deep phyla of animals can also be found in Brusca and Brusca (2003). A discussion of the history of the triune brain idea is in Pessoa and Hof (2015). The original paper by Gould and Lewontin (1979) has been republished in many venues. A good paper on sneezing sponges is Ludeman et al. (2014). A detailed comparison of invertebrate and vertebrate neurons and connections can be found in Sanes and Zipursky (2010).

Brusca, R. C., and G. J Brusca. 2003. *Invertebrates.* Sinauer.

Eitel, Michael, Hans-Jürgen Osigus, Rob DeSalle, and Bernd Schierwater. 2013. "Global Diversity of the Placozoa." *PLoS One* 8, no. 4: e57131.

Gould, Stephen Jay, and Richard C. Lewontin. 1979. "The Spandrels of San Marco and the Panglossian Paradigm: A Critique of the Adaptationist Programme." *Proceedings of the Royal Society of London,* Ser. B, 205, no. 1161: 581–598.

Ludeman, Danielle A., et al. 2014. "Evolutionary Origins of Sensation in Metazoans: Functional Evidence for a New Sensory Organ in Sponges." *BMC Evolutionary Biology* 14, no. 1: 3.

Pessoa, Luiz, and Patrick R. Hof. 2015. "From Paul Broca's Great Limbic Lobe to the Limbic System." *Journal of Comparative Neurology* 523, no. 17: 2495–2500.

Sanes, Joshua R., and S. Lawrence Zipursky. 2010. "Design Principles of Insect and Vertebrate Visual Systems." *Neuron* 66, no. 1: 15–36.

## CHAPTER 3 | THE MONKEY'S UNCULUS

A nice historical treatment of the homunculus can be found in Schott (1993), and the hermunculus is discussed by Di Noto et al. (2013). Griggs (1988) discusses the historical aspects of Mrs. Cantlie's homunculus drawings. Catania (2012) is the author of the star-nosed mole studies, and the movie of this mole foraging is at http://www.pnas.org/content/suppl/2012/06/20/1201885109.DCSupplemental/sm02.mov. Catania (Catania and Remple, 2002) is also the author of the naked mole-rat study. The naked mole-rat's visual acuity is discussed in Kott et al. (2010). The work from the O'Leary lab is published in Zembrzycki et al. (2013). Gould and Lewontin (1979) make a reprise in this chapter with their ideas about the spandrels. Leeuwenhoek's remark about ommatidia is quoted in Resh and Cardé (2009). Jessel's lecture on developing a Cyclops can be found at http://www.bing.com/videos/search?q=Thomas+Jessel+HHM+I+lecture&view=detail&mid=2E96997199F124D3005B2E96997199F124D3005B&FORM=VIRE.

Catania, Kenneth C. 2012. "Evolution of Brains and Behavior for Optimal Foraging: A Tale of Two Predators." *Proceedings of the National Academy of Sciences* 109, suppl. 1: 10701–10708.

Catania, Kenneth C., and Michael S. Remple. 2002. "Somatosensory Cortex Dominated by the Representation of Teeth in the Naked Mole-Rat brain." *Proceedings of the National Academy of Sciences* 99, no. 8: 5692–5697.

Di Noto, Paula M., Leorra Newman, Shelley Wall, and Gillian Einstein. 2013. "The Hermunculus: What Is Known about the Representation of the Female Body in the Brain?" *Cerebral Cortex* 23, no. 5: 1005–1013.

Gould, Stephen Jay, and Richard C. Lewontin. 1979. "The Spandrels of San Marco and the Panglossian Paradigm: A Critique of the Adaptationist Programme." *Proceedings of the Royal Society of London*, Ser. B, 205, no. 1161: 581–598.

Griggs, Richard A. 1988. "Who Is Mrs. Cantlie and Why Are They Doing Those Terrible Things to Her Homunculi?" *Teaching of Psychology* 15, no. 2: 105–106.

Kott, Ondřej, Radim Šumbera, and Pavel Němec. 2010. "Light Perception in Two Strictly Subterranean Rodents: Life in the Dark or Blue?" *PloS One* 5, no. 7: e11810.

Resh, Vincent H., and Ring T. Cardé, eds. 2009. *Encyclopedia of Insects*. Academic Press.

Schott, Geoffrey D. 1993. "Penfield's Homunculus: A Note on Cerebral Cartography." *Journal of Neurology, Neurosurgery and Psychiatry* 56, no. 4: 329–333.

Zembrzycki, Andreas, et al. 2013. "Sensory Cortex Limits Cortical Maps and Drives Top-Down Plasticity in Thalamocortical Circuits." *Nature Neuroscience* 16, no. 8: 1060–1067.

## CHAPTER 4 | A MATTER OF TASTE (AND ODORANT) RECEPTORS

The inner neural workings of taste reception are discussed in Simon et al. (2006). Insect taste receptors are described in Isono and Morita (2010). Bitter taste reception genes in animals is the focus of a paper by Li and Zhang (2014), and Feng et al. (2014) examine cetacean taste receptors. Bird taste receptors are discussed in Wang and Zhao (2015), and Zhao et al. (2015) focus specifically on penguin taste receptors. The *Drosophila* olfactory work is described in Woodard et al. (1989) and the Nobel Prize–winning animal olfactory work is covered in Buck and Axel (1991). Odorant receptor numbers in animals are discussed in Vandewege et al. (2016). Fabre's (1919) book on caterpillars is also mentioned. Bushdid et al. (2014) report the work on the massive number of odors we could potentially differentiate, and controversy about the estimate can be found in Gerkin (2015).

Buck, Linda, and Richard Axel. 1991. "A Novel Multigene Family May Encode Odorant Receptors: A Molecular Basis for Odor Recognition." *Cell* 65, no. 1: 175–187.

Bushdid, Caroline, Marcelo O. Magnasco, Leslie B. Vosshall, and Andreas Keller. 2014. "Humans Can Discriminate More Than 1 Trillion Olfactory Stimuli." *Science* 343, no. 6177: 1370–1372.

Fabre, Jean-Henri. 1919. *The Life of the Caterpillar.* Dodd, Mead.

Feng, Ping, et al. 2014. "Massive Losses of Taste Receptor Genes in Toothed and Baleen Whales." *Genome Biology and Evolution* 6, no. 6: 1254–1265.

Gerkin, Richard C., and Jason B. Castro. 2015. "The Number of Olfactory Stimuli That Humans Can Discriminate Is Still Known." *Elife* 4: e08127.

Isono, Kunio, and Hiromi Morita. 2010. "Molecular and Cellular Designs of Insect Taste Receptor System." *Frontiers in Neuroscience* 4: 1–16.

Li, Diyan, and Jianzhi Zhang. 2014. "Diet Shapes the Evolution of the Vertebrate Bitter Taste Receptor Gene Repertoire." *Molecular Biology and Evolution* 31, no. 2: 303–309.

Simon, Sidney A., Ivan E. de Araujo, Ranier Gutierrez, and Miguel A. L. Nicolelis. 2006. "The Neural Mechanisms of Gustation: A Distributed Processing Code." *Nature Reviews Neuroscience* 7, no. 11: 890–901.

Wang, Kai, and Huabin Zhao. 2015. "Birds Generally Carry a Small Repertoire of Bitter Taste Receptor Genes." *Genome Biology and Evolution* 7, no. 9: 2705–2715.

Woodard, Craig, et al. 1989. "Genetic Analysis of Olfactory Behavior in Drosophila: A New Screen Yields the ota Mutants." *Genetics* 123, no. 2: 315–326.

Vandewege, Michael W., et al. 2016. "Contrasting Patterns of Evolutionary Diversification in the Olfactory Repertoires of Reptile and Bird Genomes." *Genome Biology and Evolution* 8, no. 3: 470–480.

Zhao, Huabin, Jianwen Li, and Jianzhi Zhang. 2015. "Molecular Evidence for the Loss of Three Basic Tastes in Penguins." *Current Biology* 25, no. 4: R141–R142.

## CHAPTER 5 | ALL EARS (AND EYES)

Hearing in insects is reviewed in Göpfert and Hennig (2015). Billet et al. (2012) describe the vestigial status of the vestibular system in sloths, and Jerry Coyne's discussion of this phenomenon can be found at https://whyevolutionistrue.wordpress .com/2012/08/05/darwin-right-again-the-inner-ears-of-sloths-are-highly-variable/. Ekdale (2016) reviews the structure of the mammalian inner ear, and Schneider et al. (2016) review electroreception in monotremes.

Billet, Guillaume, et al. 2012. "High Morphological Variation of Vestibular System Accompanies Slow and Infrequent Locomotion in Three-Toed Sloths." *Proceedings of the Royal Society of London B: Biological Sciences* 279, no. 1744: 3932–3939.

Ekdale, Eric G. 2016. "Form and Function of the Mammalian Inner Ear." *Journal of Anatomy* 228, no. 2: 324–337.

Göpfert, Martin C., and R. Matthias Hennig. 2016. "Hearing in Insects." *Annual Review of Entomology* 61: 257–276.

Schneider, Eve R., Elena O. Gracheva, and Slav N. Bagriantsev. 2016. "Evolutionary Specialization of Tactile Perception in Vertebrates." *Physiology* 31, no. 3: 193–200.

## CHAPTER 6 | SUPERSMELLERS AND SUPERTASTERS

The description of Joy Milne's unique capacity to smell Parkinson's is detailed in Morgan (2016). Devanand et al. (2015) and Growdon et al. (2015) report on the use of diminished olfaction as a diagnostic of Alzheimer's disorder. Turin et al. (2015) and Block et al. (2015) present the arguments for the plausibility and implausibility of the vibrational hypothesis. Bloomquist et al. (2015) report on the similarity of the development of teeth and taste buds. Supertasters and fungiform papillae number was examined by Fischer et al., 2013. Cortright et al. (2007) provide an excellent review of TRP receptors and how pain is related to the receptors, and Cox et al. (2006) and Sawal et al. (2016) discuss the relation of SCN9A and nonresponse to pain. Cemre Candar's antics can be viewed at https://www.youtube.com/channel/UCTYPI3osBN9nGTSbYGmorIQ.

Block, Eric, et al. 2015. "Implausibility of the Vibrational Theory of Olfaction." *Proceedings of the National Academy of Sciences* 112, no. 21: E2766–E2774.

Bloomquist, Ryan F., et al. 2015. "Coevolutionary Patterning of Teeth and Taste Buds." *Proceedings of the National Academy of Sciences* 112, no. 44: E5954–E5962.

Cortright, Daniel N., James E. Krause, and Daniel C. Broom. 2007. "TRP Channels and Pain." *Biochimica et Biophysica Acta (BBA)—Molecular Basis of Disease* 1772, no. 8: 978–988.

Cox, James J., et al. 2006. "An SCN9A Channelopathy Causes Congenital Inability to Experience Pain." *Nature* 444, no. 7121: 894–898.

Devanand, D. P., et al. 2015. "Olfactory Deficits Predict Cognitive Decline and Alzheimer Dementia in an Urban Community." *Neurology* 84, no. 2: 182–189.

Fischer, Mary E., Karen J. Cruickshanks, Carla R. Schubert, Alex Pinto, Ronald Klein, Nathan Pankratz, James S. Pankow, and Guan-Hua Huang. 2013. "Factors Related to Fungiform Papillae Density: The Beaver Dam Offspring Study." *Chemical Senses* 38, no. 8: 669–677.

Growdon, Matthew E., et al. 2015. "Odor Identification and Alzheimer Disease Biomarkers in Clinically Normal Elderly." *Neurology* 84, no. 21: 2153–2160.

Morgan, Jules. 2016. "Joy of Super Smeller: Sebum Clues for PD Diagnostics."
   *Lancet Neurology* 15, no. 2: 138.

Sawal, H. A., et al. 2016. "Biallelic Truncating SCN9A Mutation Identified in
   Four Families with Congenital Insensitivity to Pain from Pakistan." *Clinical
   Genetics* 90, no. 6: 563–565.

Turin, Luca, Simon Gane, Dimitris Georganakis, Klio Maniati, and Efthimios M. C.
   Skoulakis. 2015. "Plausibility of the Vibrational Theory of Olfaction." *Proceedings of the National Academy of Sciences* 112, no. 25: E3154–E3154.

## CHAPTER 7 | WHERE AM I?

Muller et al. (2016) discuss the filtering out of Brownian motion in balance. For
a full discussion of balance disorders, see https://www.nidcd.nih.gov/health/bal
ance-disorders. For a detailed explanation of spinning rooms after drinking wine
especially, see Tattersall and DeSalle (2015).

Muller, Mees, Kier Heeck, and Coen P. H. Elemans. 2016. "Semicircular Canals
   Circumvent Brownian Motion Overload of Mechanoreceptor Hair Cells."
   *PLoS One* 11, no. 7: e0159427.

Tattersall, Ian, and Rob DeSalle. 2015. *A Natural History of Wine.* Yale University
   Press.

## CHAPTER 8 | TOUCHY FEELY

The medley and menagerie of touch sensor cells are discussed in Lumpkin et al.
(2010) and Connor (2006). Gregoire (2015) interviewed David Linden in 2015,
and much of the discussion of his work can be found in that publication. Lupski et
al. (2013) report on the genomic examination of Charcot-Marie-Tooth syndrome,
and Frenzel et al. (2012) report on twin studies of the same syndrome. Mathur and
Yang (2015) have examined Usher syndrome and its associated phenotypes.

Connor, Steven. 2006. "The Menagerie of the Senses." *Senses and Society* 1, no.
   1: 9–26.

Frenzel, Henning, et al. 2012. "A Genetic Basis for Mechanosensory Traits in Humans." *PLoS Biol* 10, no. 5: e1001318.

Gregoire, Carolyn. 2015. http://www.huffingtonpost.com/2015/01/20/neuroscience
   -touch_n_6489050.html

Lumpkin, Ellen A., Kara L. Marshall, and Aislyn M. Nelson. 2010. "The Cell Biology of Touch." *Journal of Cell Biology* 191, no. 2: 237–248.

Lupski, James R., et al. 2013. "Exome Sequencing Resolves Apparent Incidental
   Findings and Reveals Further Complexity of SH3TC2 Variant Alleles Causing
   Charcot-Marie-Tooth Neuropathy." *Genome Medicine* 5, no. 6: 57.

Mathur, Pranav, and Jun Yang. 2015. "Usher Syndrome: Hearing Loss, Retinal
   Degeneration and Associated Abnormalities." *Biochimica et Biophysica Acta
   (BBA)—Molecular Basis of Disease* 1852, no. 3: 406–420.

Visual acuity in baseball players is discussed in Bahill et al. (2005). Shichida and Matsuyama (2009) and Porter et al. (2012) discuss the evolution of opsins in vertebrates, and Surridge et al. (2003) focus on primate opsin evolution. A review of unconventional color vision can be found in Marshall and Arikawa (2014). Anomalous trichromacy is examined in Jordan et al. (2010), and tetrachromacy and artistic ability are addressed in Jameson et al. (2016). Neanderthal opsins are discussed in Taylor and Reimchen (2016).

Bahill, A. Terry, David G. Baldwin, and Jayendran Venkateswaran. 2005. "Predicting a Baseball's Path." *American Scientist* 93, no. 3: 218–225.

Jameson, Kimberly A., Alissa D. Winkler, and Keith Goldfarb. 2016. "Art, Interpersonal Comparisons of Color Experience, and Potential Tetrachromacy." *Electronic Imaging* 2016, no. 16: 1–12.

Jordan, Gabriele, Samir S. Deeb, Jenny M. Bosten, and J. D. Mollon. 2010. "The Dimensionality of Color Vision in Carriers of Anomalous Trichromacy." *Journal of Vision* 10, no. 8: 1–19.

Marshall, Justin, and Kentaro Arikawa. 2014. "Unconventional Colour Vision." *Current Biology* 24, no. 24: R1150–R1154.

Porter, Megan L., et al. 2012. "Shedding New Light on Opsin Evolution." In *Proceedings of the Royal Society B: Biological Sciences,* 279, no. 1726: 3–14.

Shichida, Yoshinori, and Take Matsuyama. 2009. "Evolution of Opsins and Phototransduction." *Philosophical Transactions of the Royal Society B: Biological Sciences* 364, no. 1531: 2881–2895.

Surridge, Alison K., Daniel Osorio, and Nicholas I. Mundy. 2003. "Evolution and Selection of Trichromatic Vision in Primates." *Trends in Ecology and Evolution* 18, no. 4: 198–205.

Taylor, John S., and Thomas E. Reimchen. 2016. "Opsin Gene Repertoires in Northern Archaic Hominids." *Genome* 59, no. 8: 541–549.

## CHAPTER 10 | ACCIDENTS WILL HAPPEN

Carl Sagan's (1980) classic book and de Schotten et al.'s (2015) research paper on brain imaging are references at the beginning of this chapter. Schofield et al. (2014) and Frasnelli et al. (2016) summarize the impact of TBI on olfaction, and Xydakis et al. (2015) examine TBI and olfactory disruption in U.S. servicemen. American troops were the subject of Theodoroff et al.'s (2015) meta-analysis of tinnitus and TBI in servicemen. Gourévitch et al. (2014) review the impact of everyday sounds on our hearing. Noreña and Eggermont (2006) discuss the use of acoustic treatments for tinnitus. The table describing H.M.'s response to odors can be found in Eichenbaum et al. (1983).

de Schotten, M. Thiebaut, et al. 2015. "From Phineas Gage and Monsieur Leborgne to HM: Revisiting Disconnection Syndromes." *Cerebral Cortex* 25, no. 12: 4812–4827.

Eichenbaum, Howard, Thomas H. Morton, Harry Potter, and Suzanne Corkin. 1983. "Selective Olfactory Deficits in Case HM." *Brain* 106, no. 2: 459–472.

Frasnelli, J., et al. 2016. "Olfactory Function in Acute Traumatic Brain Injury." *Clinical Neurology and Neurosurgery* 140: 68–72.

Gourévitch, Boris, Jean-Marc Edeline, Florian Occelli, and Jos J. Eggermont. 2014. "Is the Din Really Harmless? Long-Term Effects of Non-Traumatic Noise on the Adult Auditory System." *Nature Reviews Neuroscience* 15, no. 7: 483–491.

Noreña, Arnaud J., and Jos J. Eggermont. 2006. "Enriched Acoustic Environment after Noise Trauma Abolishes Neural Signs of Tinnitus." *Neuroreport* 17, no. 6: 559–563.

Sagan, Carl. 1980. *Broca's Brain: Reflections on the Romance of Science*. Presidio Press.

Schofield, Peter William, Tammie Maree Moore, and Andrew Gardner. 2014. "Traumatic Brain Injury and Olfaction: A Systematic Review." *Frontiers in Neurology* 5: 5.

Theodoroff, Sarah M., et al. 2015. "Hearing Impairment and Tinnitus: Prevalence, Risk Factors, and Outcomes in US Service Members and Veterans Deployed to the Iraq and Afghanistan Wars." *Epidemiologic Reviews* 37, no. 1: 71–85.

Xydakis, Michael S., et al. 2015. "Olfactory Impairment and Traumatic Brain Injury in Blast-Injured Combat Troops: A Cohort Study." *Neurology* 84, no. 15: 1559–1567.

## CHAPTER 11 | MODERN LIFE, STROKES, AND THE SENSES

Gourévitch et al. (2014) discuss Leq in detail, and their publication provides the basis for fig. 11.1. Noreña et al. (2006) discuss the potential of enriched acoustic environment to decrease tinnitus.

Gourévitch, Boris, Jean-Marc Edeline, Florian Occelli, and Jos J. Eggermont. 2014. "Is the Din Really Harmless? Long-Term Effects of Non-Traumatic Noise on the Adult Auditory System." *Nature Reviews Neuroscience* 15, no. 7: 483–491.

Noreña, Arnaud J., and Jos J. Eggermont. 2006. "Enriched Acoustic Environment after Noise Trauma Abolishes Neural Signs of Tinnitus." *Neuroreport* 17, no. 6: 559–563.

## CHAPTER 12 | FULL/HALF/SPLIT BRAINS

The long-term outcomes of hemispherectomies in children are reported by Moosa et al. (2013). Videos of Jack and Byron being interviewed are available at https://www.youtube.com/watch?v=f2IiMEbMnPM&list=PLtUz1usdFEouU_PwD1WNsEouO6NJIvOug and https://www.youtube.com/watch?v=3oef68YabDo, respectively. An excellent review of split brain research can be found in Gazzaniga

(2005). Ingalhalikar et al. (2014) discuss the sex differences of the wiring of the brain. The "Mike or me" experiments are described in Turk et al. (2002). The Capgras illusion is described in Hirstein and Ramachandran (1997). AgCC and its impact on language is discussed in Owen et al. (2013) and Hinkley et al. (2016). Video of Peek can be found at http://www.bing.com/videos/search?q=video+lua rence+kim+peek&view=detail&mid=4AC92FDD34CAEF953EE84AC92FD D34CAEF953EE8&FORM=VIRE.

Gazzaniga, Michael S. 2005. "Forty-Five Years of Split-Brain Research and Still Going Strong." *Nature Reviews Neuroscience* 6, no. 8: 653–659.

Hinkley, Leighton B. N., et al. 2016. "The Contribution of the Corpus Callosum to Language Lateralization." *Journal of Neuroscience* 36, no. 16: 4522–4533.

Hirstein, William, and Vilayanur S. Ramachandran. 1997. "Capgras Syndrome: A Novel Probe for Understanding the Neural Representation of the Identity and Familiarity of Persons." *Proceedings of the Royal Society of London B: Biological Sciences* 264, no. 1380: 437–444.

Ingalhalikar, Madhura, et al. 2014. "Sex Differences in the Structural Connectome of the Human Brain." *Proceedings of the National Academy of Sciences* 111, no. 2: 823–828.

Moosa, Ahsan N. V., et al. 2013. "Long-Term Functional Outcomes and Their Predictors after Hemispherectomy in 115 Children." *Epilepsia* 54, no. 10: 1771–1779.

Owen, Julia P., et al. 2013. "The Structural Connectome of the Human Brain in Agenesis of the Corpus Callosum." *Neuroimage* 70: 340–355.

Turk, David J., et al. 2002. "Mike or Me? Self-Recognition in a Split-Brain Patient." *Nature Neuroscience* 5, no. 9: 841–842.

## CHAPTER 13 | TEAM OF RIVALS MEETS THE KLUGE

Books by Marcus (2009) and Eagleman (2013) are listed below. The original Klüver-Bucy experiments can be found in Klüver and Bucy (1939).

Eagleman, David. 2012. Incognito. Robert Laffont.

Klüver, Heinrich, and Paul C. Bucy. 1939. "Preliminary Analysis of Functions of the Temporal Lobes in Monkeys." *Archives of Neurology and Psychiatry* 42, no. 6: 979–1000.

Marcus, Gary. 2009. *Kluge: The Haphazard Evolution of the Human Mind.* Houghton Mifflin Harcourt.

## CHAPTER 14 | NEURAL DETRITUS

Binocular rivalry is discussed in Levelt (1965), auditory rivalry is described in Deutsch (1974), and olfactory rivalry is discussed in Zhou and Chen (2009). Linn and Petersen (1985) describe the rod and frame test and sex differences in spatial acuity. Shams et al. (2002) report on inventing the visual illusion with sound, and

Cecere et al. (2015) expand on its workings, while Kerlin and Shapiro (2015) comment on the latter study and discuss the "Bayesian brain." The role of gloss in visual perception is examined in Adams et al. (2016).

Adams, Wendy J., Iona S. Kerrigan, and Erich W. Graf. 2016. "Touch Influences Perceived Gloss." *Scientific Reports* 6: doi:10.1038/srep21866.

Cecere, Roberto, Geraint Rees, and Vincenzo Romei. 2015. "Individual Differences in Alpha Frequency Drive Crossmodal Illusory Perception." *Current Biology* 25, no. 2: 231–235.

Deutsch, Diana. 1974. "An Auditory Illusion." *Journal of the Acoustical Society of America* 55, no. S1: S18–S19.

Kerlin, Jess R., and Kimron L. Shapiro. 2015. "Multisensory Integration: How Sound Alters Sight." *Current Biology* 25, no. 2: R76–R77.

Levelt, Willem J. M. 1965. "On Binocular Rivalry." PhD diss., Van Gorcum Assen.

Linn, Marcia C., and Anne C. Petersen. 1985. "Emergence and Characterization of Sex Differences in Spatial Ability: A Meta-Analysis." *Child Development* 56, no. 6: 1479–1498.

Shams, Ladan, Yukiyasu Kamitani, and Shinsuke Shimojo. 2002. "Visual Illusion Induced by Sound." *Cognitive Brain Research* 14, no. 1: 147–152.

Zhou, Wen, and Denise Chen. 2009. "Binaral Rivalry between the Nostrils and in the Cortex." *Current Biology* 19, no. 18: 1561–1565.

## CHAPTER 15 | PANI CA' MEUSA, CRÈME BRÛLÉE, AND SYNESTHESIA

Several papers from Charles Spence and his colleagues are discussed in this chapter —Gallace et al. (2011), Ngo et al. (2011), Spence and Gallace (2011), Knöferle and Spence (2012), Carvalho et al. (2016), and Wang and Spence (2016). Sean Day maintains a scholarly website on synesthesia at http://www.daysyn.com/. Fretwell (2011) discusses synesthesia and eugenics. Bosley and Eagleman (2015) report on twin studies of synesthesia, and Graham et al. (2014) discuss GWAS and synesthesia. The meta-analysis of imaging studies is found in Hupé and Dojat (2015). Ferrari et al. (2003) and Caggiano et al. (2009) describe mirror neuron research. Ramachandran and Brang (2008) report on tactile emotional synesthesia.

Bosley, Hannah G., and David M. Eagleman. 2015. "Synesthesia in Twins: Incomplete Concordance in Monozygotes Suggests Extragenic Factors." *Behavioural Brain Research* 286: 93–96.

Caggiano, Vittorio, et al. 2009. "Mirror Neurons Differentially Encode the Peripersonal and Extrapersonal Space of Monkeys." *Science* 324, no. 5925: 403–406.

Carvalho, Felipe Reinoso, Qian Janice Wang, Raymond Van Ee, and Charles Spence. 2016. "The Influence of Soundscapes on the Perception and Evaluation of Beers." *Food Quality and Preference* 52: 32–41.

Ferrari, Pier Francesco, Vittorio Gallese, Giacomo Rizzolatti, and Leonardo Fogassi. 2003. "Mirror Neurons Responding to the Observation of Ingestive and Communicative Mouth Actions in the Monkey Ventral Premotor Cortex." *European Journal of Neuroscience* 17, no. 8: 1703–1714.

Fretwell, Erica. 2011. "Senses of Belonging: The Synaesthetics of Citizenship in American Literature, 1862–1903." PhD diss., Duke University.

Gallace, Alberto, Erica Boschin, and Charles Spence. 2011. "On the taste of 'Bouba' and 'Kiki': An Exploration of Word-Food Associations in Neurologically Normal Participants." *Cognitive Neuroscience* 2, no. 1: 34–46.

Graham, Sarah A., et al. 2014. "Decoding the Genetics of Synaesthesia Using State-of-the-Art Genomics." Paper presented at symposium, "Synaesthesia in Perspective: Development, Networks, and Multisensory Processing," Universitätsklinikum Hamburg-Eppendorf, February 28–March 1.

Hupé, Jean-Michel, and Michel Dojat. 2015. "A Critical Review of the Neuroimaging Literature on Synesthesia." *Frontiers in Human Neuroscience* 9: 103.

Knöferle, Klemens, and Charles Spence. 2012. "Crossmodal Correspondences between Sounds and Tastes." *Psychonomic Bulletin and Review* 19, no. 6: 992–1006.

Ngo, Mary Kim, Reeva Misra, and Charles Spence. 2011. "Assessing the Shapes and Speech Sounds That People Associate with Chocolate Samples Varying in Cocoa Content." *Food Quality and Preference* 22, no. 6: 567–572.

Ramachandran, Vilayanur S., and David Brang. 2008. "Tactile-Emotion Synesthesia." *Neurocase* 14, no. 5: 390–399.

Spence, Charles, and Alberto Gallace. 2011. "Tasting Shapes and Words." *Food Quality and Preference* 22, no. 3: 290–295.

Wang, Qian Janice, and Charles Spence. 2016. "'Striking a Sour Note': Assessing the Influence of Consonant and Dissonant Music on Taste Perception." *Multisensory Research* 29, no. 1–3: 195–208.

CHAPTER 16 | CONNECTOMES

The match-to-do tests using the sea lion Rio are reported in Schusterman and Kastak (1998), and African lion individual recognition is discussed in Gilfillan et al. (2016). Ludwig et al. (2011) describe chimpanzee pitch behavior, and Deroy and Spence (2013) discuss the results of that study. The original *C. elegans* neural map can be found in White et al. (1986), and Izquierdo and Beer (2016) discuss the connectome of this nematode. Van den Heuvel et al. (2016) review comparative connectomes, and Hecht et al. (2012) discuss DTI imaging and the connectome.

Deroy, Ophelia, and Charles Spence. 2013. "Training, Hypnosis, and Drugs: Artificial Synaesthesia, or Artificial Paradises?" *Frontiers in Psychology* 4: doi: 10.3389/fpsyg.2013.00660.

Gilfillan, Geoffrey, Jessica Vitale, John Weldon McNutt, and Karen McComb. 2016. "Cross-Modal Individual Recognition in Wild African Lions." *Biology Letters* 12, no. 8: 20160323.

Hecht, Erin E., et al. 2012. "Process versus Product in Social Learning: Comparative Diffusion Tensor Imaging of Neural Systems for Action Execution—Observation Matching in Macaques, Chimpanzees, and Humans." *Cerebral Cortex* 23, no. 5: 1014–1024.

Izquierdo, Eduardo J., and Randall D. Beer. 2016. "The Whole Worm: Brain-Body-Environment Models of *C. elegans*." *Current Opinion in Neurobiology* 40: 23–30.

Ludwig, Vera U., Ikuma Adachi, and Tetsuro Matsuzawa. 2011. "Visuoauditory Mappings between High Luminance and High Pitch Are Shared by Chimpanzees (*Pan troglodytes*) and Humans." *Proceedings of the National Academy of Sciences* 108, no. 51: 20661–20665.

Schusterman, Ronald J., and David Kastak. 1998. "Functional Equivalence in a California Sea Lion: Relevance to Animal Social and Communicative Interactions." *Animal Behaviour* 55, no. 5: 1087–1095.

Van den Heuvel, Martijn P., Edward T. Bullmore, and Olaf Sporns. 2016. "Comparative Connectomics." *Trends in Cognitive Sciences* 20, no. 5: 345–361.

White, John G., Eileen Southgate, J. Nichol Thomson, and Sydney Brenner. 1986. "The Structure of the Nervous System of the Nematode *Caenorhabditis elegans*." *Philosophical Transactions of the Royal Society of London B: Biological Sciences* 314, no. 1165: 1–340.

## CHAPTER 17 | FACES AND HALLUCINATIONS

Taubert and Parr (2012) use Mooney objects to examine chimpanzee facial recognition. Dog rear Jesus can be viewed at http://dangerousminds.net/comments/jesus_christ_spotted_on_dogs_butt. Biello (2007) discusses the "God helmet," and Oliver Sacks's book on hallucinations is also listed below. Ffytche (2005) describes Bonnet syndrome, and he discusses the taxonomy of hallucination in ffytche (2013). Allen et al. (2008) use neuroimaging to study hallucinations. Blakemore et al. (2000) explain why you can't tickle yourself, and Carhart-Harris et al. (2016) discuss fMRI studies of people on LSD.

Allen, Paul, Frank Larøi, Philip K. McGuire, and Andrè Aleman. 2008. "The Hallucinating Brain: A Review of Structural and Functional Neuroimaging Studies of Hallucinations." *Neuroscience and Biobehavioral Reviews* 32, no. 1: 175–191.

Biello, David. 2007. "Searching for God in the Brain." *Scientific American Mind* 18, no. 5: 38–45.

Blakemore, Sarah-Jayne, Daniel Wolpert, and Chris Frith. 2000. "Why Can't You Tickle Yourself?" *NeuroReport* 11: 11–16.

Carhart-Harris, et al. 2016. "Neural Correlates of the LSD Experience Revealed by Multimodal Neuroimaging." *Proceedings of the National Academy of Sciences* 113, no. 17: 4853–4858.

ffytche, Dominic H. 2005. "Visual Hallucinations and the Charles Bonnet Syndrome." *Current Psychiatry Reports* 7, no. 3: 168–179.

———. 2013. "The Hallucinating Brain: Neurobiological Insights into the Nature of Hallucinations." In *Hallucination,* ed. Fiona Macpherson and Dimitris Platchias, 45–64. MIT Press.

Sacks, Oliver. 2012. *Hallucinations.* Pan Macmillan.

Taubert, Jessica, and Lisa A. Parr. 2012. "The Perception of Two-Tone Mooney Faces in Chimpanzees (*Pan troglodytes*)." *Cognitive Neuroscience* 3, no. 1: 21–28.

## CHAPTER 18 | BOB DYLAN'S NOBEL

Dunn (2015) provides a comprehensive analysis of the relations or phylogeny of languages. DeSalle and Tattersall (2012) describe the evolution of hominid brains. Scott et al. (2014), Hublin et al. (2015), and Gunz (2016) discuss the reconstruction of Neanderthal infant skulls. Bolhuis et al. (2014, 2015) describe merging and the evolution of language, and Lieberman (2015) critiques the ideas therein. Dahaene et al. (2015) discuss the emergence of literacy in humans. Lehrer (2008) describes Proust's inclination to neurobiology.

Bolhuis, Johan J., Ian Tattersall, Noam Chomsky, and Robert C. Berwick. 2014. "How Could Language Have Evolved?" *PLoS Biol* 12, no. 8: e1001934.

———. 2015. "Language: UG or Not to Be, That Is the Question." *PLoS Biol* 13, no. 2: e1002063.

Dehaene, Stanislas, Laurent Cohen, José Morais, and Régine Kolinsky. 2015. "Illiterate to Literate: Behavioural and Cerebral Changes Induced by Reading Acquisition." *Nature Reviews Neuroscience* 16, no. 4: 234–244.

DeSalle, Rob, and Ian Tattersall. 2012. *The Brain: Big Bangs, Behaviors, and Beliefs.* Yale University Press.

Dunn, Michael. 2015. "Language Phylogenies." In *The Routledge Handbook of Historical Linguistics,* 190–211. Routledge.

Gunz, Philipp. 2016. "Growing Up Fast, Maturing Slowly: The Evolution of a Uniquely Modern Human Pattern of Brain Development." In *Developmental Approaches to Human Evolution,* ed. Julia C. Boughner and Campbell Rolian, 261–283. John Wiley and Sons.

Hublin, Jean-Jacques, Simon Neubauer, and Philipp Gunz. 2015. "Brain Ontogeny and Life History in Pleistocene Hominins." *Philosophical Transactions of the Royal Society B: Biological Sciences* 370, no. 1663: 20140062.

Lehrer, Jonah. 2008. *Proust Was a Neuroscientist.* Houghton Mifflin Harcourt.

Lieberman, Philip. 2015. "Language Did Not Spring Forth 100,000 Years Ago." *PLoS Biol* 13, no. 2: e1002064.

Scott, Nadia, Simon Neubauer, Jean-Jacques Hublin, and Philipp Gunz. 2014. "A Shared Pattern of Postnatal Endocranial Development in Extant Hominoids." *Evolutionary Biology* 41, no. 4: 572–594.

Pinker's cheesecake remark is from his 1999 paper, and Sacks's brainworm statement is from his 2010 book. Hou et al. (2016) review the neural correlates for perfect pitch, and Seesjärvi et al. (2016) examine the genetic basis of pitch. Chanda and Levitin (2013) look at the neurobiology of music, and Peretz (2016) reviews the neurobiology of amusia. Genome studies of musical traits are discussed in Oikkonen et al. (2016), and Oikkonen and Järvelä (2014), Tan et al. (2014), and Gingras et al. (2015). Bonneville-Roussy et al. (2013), Ter Bogt et al. (2013), Clark and Giacomantonio (2015), Pantev et al. (2015), Savage et al. (2015), Schäfer et al. (2013), and Schäfer (2016) explore social correlates to music. Thoma et al. (2013) discuss the role of music in relieving stress, and Sievers et al. (2013) describe the Mr. Ball experiments with music and emotion. Patrick Cavanagh's art and neuroscience paper (2005) is also listed below. Huang (2009) discusses Ramachandran's take on art. Colton (2012) describes Painting Fool, and Berov and Kühnberger (2016) discuss the hallucinogenic capacity of Painting Fool.

Berov, Leonid, and Kai–Uwe Kühnberger. 2016. "Visual Hallucination for Computational Creation." In *Proceedings of the Seventh International Conference on Computational Creativity,* 107–114.

Bonneville-Roussy, Arielle, Peter J. Rentfrow, Man K. Xu, and Jeff Potter. 2013. "Music through the Ages: Trends in Musical Engagement and Preferences from Adolescence through Middle Adulthood." *Journal of Personality and Social Psychology* 105, no. 4: 703.

Cavanagh, Patrick. 2005. "The Artist as Neuroscientist." *Nature* 434, no. 7031: 301–307.

Chanda, Mona Lisa, and Daniel J. Levitin. 2013. "The Neurochemistry of Music." *Trends in Cognitive Sciences* 17, no. 4: 179–193.

Clark, Shannon Scott, and S. Giac Giacomantonio. 2015. "Toward Predicting Prosocial Behavior: Music Preference and Empathy Differences between Adolescents and Adults." *Empirical Musicology Review* 10, no. 1–2: 50–65.

Colton, Simon. 2012. "The Painting Fool: Stories from Building an Automated Painter." In *Computers and Creativity,* 3–38. Springer.

Gingras, Bruno, et al. 2015. "Defining the Biological Bases of Individual Differences in Musicality." *Philosophical Transactions of the Royal Society B: Biological Sciences* 370, no. 1664: 20140092.

Hou, Jiancheng, et al. 2016. "Neural Correlates of Absolute Pitch: A Review." *Musicae Scientia* doi 10.1177/1029864916662903.

Huang, Mengfei. 2009. "The Neuroscience of Art." *Stanford Journal of Neuroscience* 2, no. 1: 24–26.

Oikkonen, Jaana, and Irma Järvelä. 2014. "Genomics Approaches to Study Musical Aptitude." *Bioessays* 36, no. 11: 1102–1108.

Oikkonen, Jaana, Päivi Onkamo, Irma Järvelä, and Chakravarthi Kanduri. 2016.

"Convergent Evidence for the Molecular Basis of Musical Traits." *Scientific Reports* 6: 39707.

Pantev, Christo, et al. 2015. "Musical Expertise Is Related to Neuroplastic Changes of Multisensory Nature within the Auditory Cortex." *European Journal of Neuroscience* 41, no. 5: 709–717.

Peretz, Isabelle. 2016. "Neurobiology of Congenital Amusia." *Trends in Cognitive Sciences* 20, no. 11: 857–867.

Pinker, Steven. 1999. "How the Mind Works." *Annals of the New York Academy of Sciences* 882, no. 1: 119–127.

Sacks, Oliver. 2010. *Musicophilia: Tales of Music and the Brain.* Vintage.

Savage, Patrick E., Steven Brown, Emi Sakai, and Thomas E. Currie. 2015. "Statistical Universals Reveal the Structures and Functions of Human Music." *Proceedings of the National Academy of Sciences* 112, no. 29: 8987–8992.

Schäfer, Thomas. 2016. "The Goals and Effects of Music Listening and Their Relationship to the Strength of Music Preference." *PloS One* 11, no. 3: e0151634.

Schäfer, Thomas, Peter Sedlmeier, Christine Städtler, and David Huron. 2013. "The Psychological Functions of Music Listening." *Frontiers in Psychology* 4: 511.

Seesjärvi, Erik, et al. 2016. "The Nature and Nurture of Melody: A Twin Study of Musical Pitch and Rhythm Perception." *Behavior Genetics* 46, no. 4: 506–515.

Sievers, Beau, Larry Polansky, Michael Casey, and Thalia Wheatley. 2013. "Music and Movement Share a Dynamic Structure That Supports Universal Expressions of Emotion." *Proceedings of the National Academy of Sciences* 110, no. 1: 70–75.

Tan, Yi Ting, Gary E. McPherson, Isabelle Peretz, Samuel F. Berkovic, and Sarah J. Wilson. 2014. "The Genetic Basis of Music Ability." *Frontiers in Psychology* 5: 658.

Ter Bogt, Tom F. M., Loes Keijsers, and Wim H. J. Meeus. 2013. "Early Adolescent Music Preferences and Minor Delinquency." *Pediatrics* 131, no. 2: e380–e389.

Thoma, Myriam V., et al. 2013. "The Effect of Music on the Human Stress Response." *PloS One* 8, no. 8: e70156.

## CHAPTER 20 | NO LIMITS

The study of reading comprehension on the computer screen versus hard copy was accomplished by Mangen et al. (2013). Ortiz De Gortari (2016) explains the Game Transfer Phenomena (GTP) in detail. The impact on tactile and sensory aspects of perception and homuncular flexibility in a virtual reality world is discussed in Won et al. (2015). My colleague and I discuss Antonio Damasio's views on emotions in DeSalle and Tattersall (2012), and Tattersall (2016) discusses the role of language in human cognition.

DeSalle, Rob, and Ian Tattersall. 2012. *The Brain: Big Bangs, Behaviors, and Beliefs.* Yale University Press.

Mangen, Anne, Bente R. Walgermo, and Kolbjørn Brønnick. 2013. "Reading Linear Texts on Paper versus Computer Screen: Effects on Reading Comprehension." *International Journal of Educational Research* 58: 61–68.

Ortiz De Gortari, Angelica B. 2016. "The Game Transfer Phenomena Framework: Investigating Altered Perceptions, Automatic Mental Processes and Behaviors Induced by Virtual Immersion." *Annual Review of Cybertherapy and Telemedicine* 14: 9–15.

Tattersall, Ian. 2016. "The Thinking Primate: Establishing a Context for the Emergence of Modern Human Cognition." *Proceedings of the American Philosophical Society* 160, no. 3: 254–265.

Won, Andrea Stevenson, Jeremy Bailenson, Jimmy Lee, and Jaron Lanier. 2015. "Homuncular Flexibility in Virtual Reality." *Journal of Computer-Mediated Communication* 20, no. 3: 241–259.

# INDEX

Italic page numbers indicate figures.

acid recognition, 76
acorn worms, 114
action potentials, 17, 75, 86, 101, 184
Adachi, Ikuma, 208
Adams, Wendy, 182
adaptation, 4, 36, 175, 207, 209–10
ADCY8 gene, 245
agenesis of the corpus callosum
   (AgCC), 158–59
ageusia, 137
aging: balance and, 90–91; hearing
   and, 91, 94; touch and, 102. *See
   also* Alzheimer's disease
agnosia, 147, 164–65
agraphia, 147
airway, 42
alexia, 147
algae, 12
*Aliivibrio fischeri,* 9
Allegri, Gregorio, 249–50

Allen, Paul, 225–26
allyl isothiocyanate, 82
alpha-synuclein, 71–72, 223
alpha waves, 179
Alzheimer's disease, 67–69, 223
American Museum of Natural History, 178
American Society of Brewing Chemists, 78
amnesia, 124
amoeba, 12
amphibians, 23
ampullae, 85, 86, 88–90
amusia (tone deafness), 98, 244, 245
amygdala, 101, 136, 157, 229
anencephaly, 148, 149
animals: opsin genes in, 118; smell
   and taste in, 41–56; tactile and
   balance senses in, 27–40. *See also
   specific types of animals*

blind spots, 32–33, 35

Block, Eric, 73

Blombos Stone, 233, *233*

blood clots, 144, *144*

blood pressure, ix, 247

Bolhuis, Johan, 237

Bosley, Hannah, 196

bottom-up approach to perception, 162, 167

braille, 232, 237

brain: anatomy of, 23, 123–27, *127*, 200–201; bilaterian invention, 17–18; cerebrum, 154; common to almost all species, 2; computer analogy to, 160–61; deformation of, 134; evolution of, 19–26, 211; of extinct specimens, 234; fetal brain development, 27, 235; gray and white matter, 214–15, *215*; infant brains, 235; "kluge" as term for, 160; lateralization of, 213; left brain as dominant side, 153–56; location of, 130–31; neomammalian, 21–22; number of cells in, 27; paleomammalian, 21–22; plasticity of, 142, *143*, 150; size of, 211, 235; split brain and split-brain experimental design, 152–58, *156*; "team of rivals" analogy to, 160–61; tripartite, 21–24, *23*, 41; vertebrate compared with insect and nematode, *19*. *See also* cerebellum; cortex

brain fart, 145

brain imaging techniques, 127, 135, 200–201, 214–16, 243. *See also specific techniques*

brain injuries, 122–38; Aubertin's experiments on Cullerier's brain, 124; Broca's autopsy of Tan's (Leborgne) brain, *125*, 126, 127, 129–30, 151; Gage's skull at Har-

vard's Library of Medicine, 124, *125*, 127, 129–30; H. M.'s (Molaison) loss of short-term memory, 124–25, *125*, 127–29, *130*; loss of speech (aphasia), 126; Mr. B's loss of short-term memory, 124; sidedness and, 150–51; Wearing's loss of short-term memory, 124, *125*. *See also* stroke; traumatic brain injury (TBI)

brain stem, 21, 101, *164*, 169

Breundl, Franz, 124

Broca, Paul, 126–27, 152

Broca's area, 126, *127*, 129, 151, 169, 215, 227, 234–35, 238

*Broca's Brain: Reflections on the Romance of Science* (Sagan), 127–28

Brown, Georgia, 93–94

Brownian motion, 86

Buck, Linda, 49, 52

Bucy, Paul, 165–66

*Bullacris membracioides*, 60

Bullmore, Edward, 213

butterflies, 53, 118

Candar, Cemre, 81–83

Cantlie, Mrs. H. P., 28–29

Capgras syndrome, 157, 219

capsaicin, 81–83

Carey, Mariah, 92–94

Carhart-Harris, Robin, 228

Carlson, John, 50

carnivores, 44–45, 207–8. *See also specific types*

carotid artery, 144

Carson, Rachel, 205

Catania, Kenneth, 30

cats: environmental advantages of crossmodality for, 207; hearing loss in, 143; night vision in, 111; taste receptors in, 43–44, 47

Cavanagh, Patrick, 251–53
CCI (controlled cortical impact), 133
Cecere, Roberto, 179
cephalochordates, 19, 20
cerebellum, 21, 41, 86, 154, 169;
    ascending pathways to, 169;
    descending pathways from, 169
cerebrum, 154
cetaceans, 47
Charcot-Marie-Tooth disorder
    (CMT), 103–5, 197
Charles Bonnet syndrome, 225
chemoreception, 168
Cheok, Adrian David, 261
chimpanzees: connectomes in, 214,
    214; crossmodality studies on,
    208–9; facial recognition by,
    218–19; infant brain development
    in, 235–36; infant brains in, 232;
    language and, 237; larynx in, 234;
    light-from-above effect and, 210;
    mirror effects in, 215–16; odorant
    receptor genes in, 54; proteins sim-
    ilar to humans, 37; sign language,
    use of, 230
Chinese Cat Boy, 111
chlorophyll, 5
chloroplast, 10–11
chocolate, crossmodality of taste and
    names for, 186–87
chordates, 19
chordotonal organs, 60
chromophores, 11, 113
chromosomes, 116, 197–98
cilia, 70, 86, 87, 89, 94, 123
cingulate gyrus, 159
circadian rhythm, 112
Clamydomonas (Chlamy), 12
clinico-anatomical correlation
    method, 124, 130, 136, 147, 164
CMT (Charcot-Marie-Tooth disorder),
    103–5, 197

cnidarians, 17, 114
cochlea. See ears
cochlear nuclei, 169
cognitive development, 235–36
color vision and color blindness, 5,
    113, 115–21, 147, 166–67, 173
Colton, Simon, 254
comb jellies. See ctenophores
computed tomography (CT) scans,
    127, 134
computers: analogy to the brain,
    160–61; daily use of, 260; evolution
    of, 259–60; personalization of, 260
concussion, 131–35; olfactory dys-
    function due to, 136
conditioning by reinforcement, 207
cones. See rods and cones
congenital brain anomalies, 148, 149,
    217
connectomes, 211–15, 214
consciousness, x, xi, 263–64
convergence, 4
cornea, 36, 109
corpus callosum, 150, 152, 155, 158–59
cortex, 21, 22–23, 23, 41; alpha wave
    used to measure involvement of,
    179; auditory, 145, 164, 169, 225,
    238; auditory cortex, 244; barrels
    in, 30, 31–32; cerebral cortex, 70,
    86; cortical thickness, 243–44;
    entorhinal cortex, 68; in fish, viii;
    gustatory cortex, 168; in mice, 32;
    in naked mole-rats, 31; occipital
    cortex and lobe, 146; olfactory cor-
    tex, 70, 170; orbitofrontal cortex,
    70, 164, 168; in primates, 215;
    retrosplenial cortex, 228; sensory
    cortex, 163, 170; somatosensory,
    101, 170; temporal cortex, 157;
    ventral occipital cortex, 165; visual
    cortex, 163–64, 166, 227–28, 238,
    238

cortico-basal ganglia-thalamic loop, 41–42
Cox, James, 83–84
Coyne, Jerry, 63
cranial nerves, 168–69
Crick, Francis, 258, 263
crossed hands deficit, 175–76, *177*, *181*
crossmodalities, x–xi, 130, 160–71, 177–84; audio-visual crossmodality, 208; cross-activation process, 201–2; cultural differences in, 190–91; environmental challenges and, 206–7; gaming and, 261; of languages, 232; lions and, 208; olfactory-music crossmodality, 191; synesthesia, x, 178, 193–201; tactile-auditory crossmodality, 181; tactile-visual crossmodality, 182; taste and other senses, 185–204; visual-auditory crossmodality, 178–79; visual-vestibular crossmodality, 182–83
crows, 48
crustaceans, 118
CT (computed tomography) scans, 127, 134
ctenophores (comb jellies), 16, 114
cubism, 253
cubozoans, 17
Cullerier, Monsieur, 125–26
cultural differences in crossmodality, 190–91
cupulae, 86–90
cyanobacteria, 10
cyborgs, 256
Cyclops phenomenon, 39–40, 59

Damasio, Antonio, 264
Darwin, Charles, 15–16, 18, 26, 59, 63
Dawkins, Richard, 222

Day, Sean, 193
deafness and hearing loss: correlation with low tactile acuity, 106; in insect ancestors, 58; restoration of hearing, xii. *See also* hearing, sense of; tinnitus
decision-making, brain lobe involved in, 168
deer, 71–72
Dehaene, Stanislas, 240
dementia, 223
dendrites, 214
Denisovan, 121, 236
depolarization, 75
dermis. *See* skin
Deroy, Ophelia, 210–11
de Schotten, Michel Thiebaut, *125*, 127–30, 234
deuteranopia, 117
deuterium approach, 73–74
deuterostomes, 17–19
Deutsch, Diana, 174
diabetes, 90, 102
dichromatic vision, 116, 120
diffusion tensor imaging (DTI), 135, 201–2, 212, 214–16
digital geometry processing, 127
diminished-sensing humans, x
*Dionaea muscipula* (Venus flytrap), *12*, *13*
discrimination tests, 54
dizygotic twins, 105
DNA, 8, 84, 103–5, *104*, 116, 121, 258–59
"dog rear Jesus," 219–20
dogs, 34, 207
Dojat, Michel, 202
dolphins, 47, 63
dopamine, 42
Doppler illusion, 202–3
dorsal root ganglia, 101
double vision, 146

drinking: beer taste, 78–79; dizziness as result of, 87–88

*Drosophila melanogaster. See* fruit fly

drug addiction, 42

drug-induced hallucinations, 222–23

DTI (diffusion tensor imaging), 135, 201–2, 212, 214–16

duck-billed platypus, 63

Dunn, Rob, 3

Dunn's Provocation, 3

Dylan, Bob, 230, 240

Eagleman, David, 160–62, 196

ear canal, 95

eardrum, 64, 95–96, 139

ears: anatomy of, 57–58, 64–65, 87, 95–96, 97, 139; cochlea, 65, 95–97, 97, 247; in fish, 61–64; hammer (malleus), 64, 95–96; inner and outer hairs in, 96; inner ear, 57, 61–63, 65, 87, 88, 89, 91, 95, 138, 139, 146, 168–69, 183, 247, 248; in insects, 60–61; middle ear, 91, 95; organ of Corti, 96–97, 169; ossicles, 95–96; outer ear, 64–65, 95; protective purpose of, 122; semicircular canals, 62, 85–86; stirrup (stapes), 95–96, 97; in vertebrates, 60–62. *See also* hearing, sense of

earworms, 242

echinoderms, 19

echolocation, 58–59

Eggermont, Jos, 143

Eimer's organs, 29

Eldredge, Niles, 59

electrolocation, 62

electromagnetic radiation, 4, 5

electroreception, 62–63

elephants, 51

11-cis-retinal, 113

embolisms, 144

embryo development, 32, 38–39

emotional responses: evolutionary history of, 264; linked to sensations, 203–4; Mr. Ball used to indicate, 250–51, 251; to music, 247, 251

endolymph, 85, 87, 89

endoplasmic reticulum, 75

engulfment events by early cells, 11

environmental perception, 4; mammals and, 206; single-celled organisms and, 205; sponges and placozoa, 16

epileptic seizures, surgery to treat, 124–25, 149, 150, 152–53

Erwin, Terry, 3

ethanol, 88

eugenics, 195

eukaryotes, 8, 11–12, 15

*Euprymna scolopes*, 9

evolution: brain and, 160–61, 213; color vision and, 114; ears and, 61–62; environmental challenges prompting, 206; gradualism and, 59; process of, 97–98; single-celled organisms and, 205; visual perception and, 179–80. *See also* adaptation; genetic drift; natural selection

exaptation, 45, 60

exceptions in nature, 1–2

expectancy violations theory (EVT), 208

explosions: hearing damage from, 140; traumatic brain injury and, 133, 135–38

eyes: anatomy of, 35, 109, 109; compound eyes, 36–37; injuries and disorders as source of hallucinations, 224; placement during embryo development, 38–40. *See also* retina; vision

Fabre, Jean-Henri, 55

face-morphing technique, 155–56

facial recognition, 217–20
false hellebore, 40
Farley, Chris, 41
feedback: sound, 140; vision, 167
Feeney weight drop, 133
Feng, Ping, 47
fetal brain development syndromes.
  *See* congenital brain anomalies
ffytche, Dominic, 225–26
field of vision, 34–35, 40, 146, 260–61
figure skaters, 86–88, 182–83
Finch, Lana, 30
finches, 48
fish, viii; balance in, 62; bones, evolution of, 61; electrolocation in, 62;
  jaws in, 64; opsin genes in, 114;
  taste receptors in, 47; in tree of life,
  22. *See also specific types*
flies: exaptation in, 60; gustatory
  receptors (GRNs) in, 46; mouthparts of, 45–46; response to hot and
  cold, viii–ix. *See also* fruit fly
flour beetle, 46
fMRI (functional magnetic resonance
  imaging), 197, 201, 202, 225–26,
  237–38
Foer, Jonathan Safran, 66
football-related injuries, 132–33
fovea, 29–30, 110–11
foveal fixation, 110–11
FoxP2 gene, 236
FPI (fluid percussion injury), 133
Franklin, Rosalind, 258
free-nerve endings, 99, 100
Fretwell, Erica, 195
frogs, 47, 114
fruit fly (*Drosophila melanogaster*),
  18, 38, 48–51, 56, 105
functional magnetic resonance imaging (fMRI), 197, 201, 202, 225–26,
  237–38
fusiform gyrus, 157, 165, 217

GABA (gamma-aminobutyric acid), 42
Gage, Phineas, 124, *125*, 127,
  129–30, 234
Galileo Galilei, 3
Galton, Francis, 195–96
galvanic vestibular stimulation (GVS),
  183
*Game of Thrones,* Dothraki language
  in, 230–31, *231*
game transfer phenomena, 261
gaming, 261
gandolium, 145
Garten, Ina, 185
Gates, Bill, 260
Gazzaniga, Michael, 152, 154, 155–57
gender: crossed hands deficit and,
  176; male vs. female brains, 153–54;
  rod-and-frame test and, 177; sex
  bias in research on homunculus, 29;
  sex chromosomes, 116; sex-linked
  color vision, 116, 118–19, 120;
  synesthesia and, 193–94; women's
  ability to lateralize, 152
gene duplication, 115
gene regulation, 38–40
genes: number of, 72, 74, 116. *See
  also specific genes*
genetic drift, 26, 161
genetics, 37, 59, 84, 98, 105–6, 173,
  242; population genetics, 250; of
  synesthesia, 195–96. *See also* DNA
genome sequencing, 48, 50–51,
  103–5, *104*, 114–15, 121, 197
genome-wide association studies,
  196–97
Gimble, Johnny, 139
glia, 134
glomeruli, 51, 69, 70
glossiness, perception of, 182
"god helmet," 221
Goldschmidt, Richard, 59
Goodwin, Doris Kearns, 160, 161

language (*continued*)
  236; relationship among, 231–32;
  ritualistic objects, 233
Language Creation Society, 230
larynx, 234
Lascaux cave drawings, 252
lateral geniculate nuclei, 163, *164*
lateral line, 62
Leborgne, Louis Victor, *125*, 126,
  127, 129–30, 151
Leeuwenhoek, Antonie van, 36
Lehrer, Jonah, 240, 251
lens of the eye, 109
Leonardo da Vinci, 97; *Last Supper*,
  254; *Mona Lisa*, 253
Lepidoptera (moths and butterflies), 53
Leq (equivalent sound level), 141, *141*
Levelt, Willem, 172
Lewontin, Richard, 24, 35
Lewy, Frederic, 223
Lewy body dementia, 223–24
Li, Diyan, 47, 48
lice, 46
*The Life of a Caterpillar* (Fabre), 55
light: bacteria and, 10; color vision
  and, 5, 113; first organisms to
  perceive, 3; how light enters the
  nervous system, 4; single-celled
  organisms detecting, 8
light-from-above effect, 210, *210*
limbic system, 21–23, 41, 157, 168,
  169, 229
Lincoln, Abraham, 160, *161*
Linden, David, 101
lingual gyrus, 165
lions, 208
literacy, 237–38, *238*, 260
lizards, 114
lock-and-key mechanism, 6, 9, 72,
  73, 123
loss of short-term memory, 124–25,
  132, 149

LSD, 222, 228
Ludwig, Vera, 208
Lunde, Darrin, 57
Lupski, James, 103–5, 197
Lyell, Charles, 59

macaques. *See* monkeys
magnetic resonance imaging (MRI),
  127, 128, 134, 145, 147, 201, 212,
  258
magnetism and magnetic field, 8–10, 14
mal de débarquement syndrome, 90
malleus, 64
mammals: balance in, 62; connectomes in, 213; environmental
  sensory input and, 206; jaws in, 64,
  64–65; opsin genes in, 115; taste in,
  48. *See also specific types*
manatees, 48
Marcus, Gary, 160
marketing: music and, 189; product's
  name and, 186
Marmarou weight drop, 133
Martin, Steve, 241
Master Brewers Association of the
  Americas, 78
match-to-do tests, 207–8, *208*
Matsuzawa, Tetsuro, 208
medulla, 169
Meeus, Wim, 247
megabats (Megachiroptera), 58
Meissner's corpuscles, 99–101, *100*
memories giving context, xi, 71, 204
memory: loss of, 124–25, 132, 149;
  vision and, 162
men. *See* gender
Meniere's disease, 89
Merkel cell-neurite complexes, 99–100
metallic taste, 136–37, 148
mice: barrels and PAX6 mutation,
  32–33; connectomes and, 213;
  neocortex in, 31; plaque formation

in brains of, 67; small eye mutation, 32; whiskers in, 31

microbats (Microchiroptera), 58–59

microbes, 205–6

microscopy, 258

Mielgaard, Morten, 78

Millet, Jules, 195

Milne, Joy, 66–67, 69–71

mimosa, 12

ministrokes, 145

mirror invariance, 239

mirror neurons, 203, 215–16, 227

"Miserere" (Allegri), 249–50

mitochondria, 75

mitral cells, 70

Mohawks as skywalkers, 89

Molaison, Henry (H. M.), 124, 125, 127–30, 130, 149, 150, 212

moles: naked mole-rats, 30–31; star-nosed, 29–30, 30

mollusks, 35

monkeys: color vision in, 120; connectomes in, 213, 214; crossmodality studies on, 208; facial recognition by, 217; homunculi in, 29; macaque brain ablation studies, 165–66; mirror effects in, 215–16; visual agnosia in, 165

monochromatic vision, 116–17, 256

monotremes, 63, 115

Mooney objects, 219

Moore, Gordon, 260

Moore's law, 260

Moosa, Ahsan, 150

Morgan, Thomas Hunt, 49

morphologies, 37, 234, 243

moths, 53, 55

mouseunculus, 30, 31–33

mouthparts, insect, 46, 60

Mr. Ball, 250–51, 251

MRI (magnetic resonance imaging), 127, 128, 134, 145, 147, 201, 212, 258

Mukherjee, Pratik, 159

Muller, Mees, 86

multicellular life, emergence of, 11–12, 15

multisensory integration. *See* cross-modalities

music, 241–51; amusia, 244, 245; earworms, 242; genetic basis of, 242–43; musical preferences, 246–49; perception of taste and, 188–89; phenotypic complexity of, 245; pitch, 243–45; stress modulated by, 250; as universal language, 249; vibration frequency, 243

musky odor, 70–73

mutation, 32–33, 161

myelination, 105

nasal bilateral rivalry, 175

nasal epithelium, 69, 70

National Health and Nutrition Examination Survey (NHANES), 90–91

National Institute of Deafness and Other Communication Disorders, 88

Native American skywalkers, 88, 89

natural selection, 24, 26, 161; balance and vestigial organs, 63; environmental advantages of crossmodality, 207; field of vision and, 34–35; music and, 244–45

Neanderthals, 25, 121, 211, 233–36

neglect, 147

nematodes, 51, 105, 212; brain compared with insects and vertebrates, 18, 19

neocortex, 22–23, 31

neomammalian brain, 21–22

neonatal synesthesia hypothesis, 199

Neubauer, Simon, 235

neural detritus, 183–84

neural fibers, 158

peppers, calibrating heat of, 82–83

perception, viii–xi, 162–63, 179, 263

perfect pitch, 98, 243–45

perilymph fistula, 90

peripheral nervous system, 123

peripheral vision, 34

Persinger, Michael, 221–22

personality and musical preference, 247–48

phantom limb syndrome, 262

pheromones, 53, 55

photons, 5, 6, 113

photosensitive retinal ganglion cells (pRGC), 112

pigeons, 34–35

Pingelap (Micronesian island), 117

pitch, 92–94, 98, 189–90, 199–200, 203, 243–47; ADCY8 gene and, 245; AVPR1A gene and, 247

PKD2L1, 43

placozoa, 16, 114

*Plant Neurobiology*, 12

plants, viii; central sensing system of, 12–13; genome duplications in, 115; neurobiology of, 12; response to light, 12; similarity with bacteria in use of light, 10–11

planum temporale, 243–44

Plato, 241

platypus, 29, 63, 115

pleasure, 42

pleiotropy, 59

Pollan, Michael, 13

Pollock, Jackson, 253

polycystic kidney disease (PKD), 43

polymorphisms, 196–98

polyphagous insects, 46

pons, 169, 223

positron emission tomography (PET) scans, 68, 201

Potter, Beatrix, 38–39

"prain," 18

primates: arboreal, 34; balance in, 62; brain evolution in, 213, 215; crossmodality responses in, 208–9; mirror neurons in, 203; opsin genes in, 115

primordial membrane, 3

*Principles of Geology* (Lyell), 59

probabilistic analysis, 180

probability theory, 26

proprioception organ, 60, 170

prosopagnosia, 217

protanopia, 117

Proteobacteria, 8

proteomammalian brain, 21–22

protocadherin, 247

protostomes, 17–19

Proust, Marcel, 70–71, 240, 251

proverbs, interpretation of, 159

pruning, 199, 200, 243, 245

pseudogenes and pseudogenization, 43, 44, 47, 48, 51, 54, 72

psychoeducation, 143

psychotherapeutic counseling, 143

pupil, 109

Pynchon, Thomas, 217

quorum sensing, 8, 9, 15, 38, 205

radio waves, 259

*Rain Man* (film), 158–59

Ramachandran, V. S., 156–57, 252, 262

rats, 29, 142

R-complex, 21

red-green color blindness, 117

Rees, Geraint, 179

Reimchen, Thomas, 121

religion, 222

religious icons, appearance of, 219–20

*Remembrance of Things Past* (Proust), 71

Remple, Fiona, 30

smell, sense of, 6, 48–56; Alzheimer's disease and, 67; compared to taste and sight, 123, 257; concussion and TBI causing loss of, 136; crossmodality of, 191–92; difference from sense of taste, 45; discrimination tests, 54; in early vertebrates, 21; in fruit flies, 48–51, 56; in humans, 53–56; in insects, 45, 53; neural pathway for, 164; number of odors, 257; olfactory hallucinations, 226; Parkinson's disease and, 66–67, 69, 71; in rats, 49; stroke and, 147–48; supersmellers, 69, 70–71; vibrational theory, 73–74

Smith, Amy, 97

Sniffin' Sticks, 136

somatosensory receptors, 99

Sonic Hedgehog (Shh), 38–39

*The Sopranos* (television show), 222

sound: characteristics of, 92; crossmodality of taste and names for food, 186–87; how sound is created, 7; intensity of, 92, 94–95; loudness of, 203; pitch, 92–94, 98, 189–90, 199–200, 203; range of in hertz, 92, 257; sound waves, 91; tone, 92. *See also* hearing, sense of

spandrels, 35–36

speech. *See* language

speech aphasia. *See* aphasia

Spence, Charles, 186–91, 210–11, 257

Sperry, Roger Wolcott, 149, 152–54

spina bifida, 148

spinal cord, 134, 169, 170

sponges, 16, 114

Sporns, Olaf, 213

sporting events and intensity of sound, 94–95

sports-related concussions, 132, 135

squid culturing *A. fischeri*, 9

starfish, 19

stereocilia, 139–40, 142

stereovision, 34–35

Stiller, Ben, 99

stirrup (stapes), 95–96, 97, 98, 138

stomatopods, 118

Storms, Tim, 94

stress, modulated by music, 250

stroke, 102, 126, 139, 142–48, 144, 204

Stroop test, 187–88, 194

substantia nigra, 223

sunflowers, 12

superior temporal gyrus, 169

super senses, x

supersmellers, 69, 70–71

supertasters, 77–80

"swipe card theory," 73

Sylwester, Robert, 1

symbolic thought, 233

synapses, 18, 20, 76, 199, 214

synesthesia, x, 178, 193–204, 236; grapheme-color synesthesia, 198–99, 201, 202, 204; intramodal, 202–3; pitch and, 199–200

TAARs (trace amine-associated receptors), 51–52

tactile fovea, 29–30

tapetum lucidum, 111

TASs, 42–43; TAS1R2 (sweet taste receptor), 43–45; TAS1R3 (sweet taste receptor), 43; TAS1s (sweet taste receptor), 42–43; TAS2s (bitter taste receptors), 43, 46–48

taste, sense of, 6, 41–48, 74–84; basic kinds of taste, 42, 75, 257; brain injury and, 136; categories of humans for taste, 77–78; compared to pain, 80–81; compared to sight, 123; compared to smell, 45, 123, 257; crossmodality of, 185–204; hallucination, 136; in insects, taste (*continued*)

45–46; loss of taste (ageusia), 137; multisensory nature of, 168; music and perception of taste, 188–89; names of products, influence of, 186; nontasters, 77–80; pani ca' meusa and perception of taste, 185–86, 187; peppers and, 82–83; sense of smell linked to, 168; stroke and, 147–48; supertasters, 77–80; taste receptors, 42–48, 44–45, 74–76, 80, 168, 257; taste wheel, 78; vibrational theory, 76. *See also* TASs

taste buds, 80, 257

Tattersall, Ian, 264

Tatum, Edward, 37

Taubert, Jessica, 218–19

Taylor, John, 121

TBR1 gene, 197–98

"team of rivals" analogy, 160–61, 188

telescopes, 259

temperature reception, viii–ix

temporal lobes, 145, 150, 157, 164–65, 217, 238

ter Bogt, Tom, 247

test of genuineness (TOG) and revised test of genuineness (TOG-R), 194

tetrachromatic individuals, 196, 254–55

tetrachromats, 118–21

thalamus, 23, 33, 136, 168, 169, 204

Theodoroff, Sarah, 137

three-toed sloths, balance in, 63

Tierney, Lawrence, 15

tinnitus, 138, 140, 143, 145

tone, 92

tone deafness, 98

tongue, 42, 61, 77, 80, 168

top-down approach to perception, 162, 167

touch, sense of, 99–107; aging and, 102; bilateral symmetry of organs for, 175; blindness and, 107; brain injury and, 136; cross-modal interaction, 163, 187; deafness correlated with low tactile acuity, 106; first and most important of all senses, 10; first responder cells and, 123; lanceolate endings, 100–101; language and, 232; Meissner's corpuscles, 99–101, *100*; Merkel cell-neurite complexes, 99–100; mirror touch, 203; multisensory integration with other senses, 170–71; Ruffini endings, 99, 100, *100*; scientific understanding of, 102; steps in process, 101; tactile hallucinations, 226; Usher syndrome and, 106–7

touch aversion, 102

touch receptor, *100*

transcranial magnetic stimulation (TMS), 143

transient receptor protein (TRP), 82

transmembrane domains, 51

trauma, effect of, x

traumatic brain injury (TBI), 135–38; concussion, 131–35; deformation of brain, 134; hearing loss related to, 137–38; military-related events and explosions as cause of, 135–38; olfactory dysfunction related to, 136; taste loss related to, 136–37; touch and skin sensitivity related to, 137; vision impairment related to, 137

tree of life: Darwin's design and, 15–16; as organizing principle, 2, 16; vertebrate brain changes and, 161

trichromatic vision, 115–16, 118–20

trilobite fossils, 36